T0221163

Carbon

CARBON

A Biography

Bernadette Bensaude-Vincent
and
Sacha Loeve

Translated by Stephen Muecke

polity

Originally published in French as *Carbone. Ses vies, ses oeuvres* © Éditions du Seuil, 2018

This English edition © Polity Press, 2024

Polity Press
65 Bridge Street
Cambridge CB2 1UR, UK

Polity Press
111 River Street
Hoboken, NJ 07030, USA

ISBN-13: 978-1-5095-5920-6 – hardback

A catalogue record for this book is available from the British Library.

Library of Congress Control Number: 2023948422

Typeset in 11.5 on 14 Adobe Garamond
by Fakenham Prepress Solutions, Fakenham, Norfolk NR21 8NL
Printed and bound in Great Britain by CPI Group (UK) Ltd, Croydon

The publisher has used its best endeavours to ensure that the URLs for external websites referred to in this book are correct and active at the time of going to press. However, the publisher has no responsibility for the websites and can make no guarantee that a site will remain live or that the content is or will remain appropriate.

Every effort has been made to trace all copyright holders, but if any have been overlooked the publisher will be pleased to include any necessary credits in any subsequent reprint or edition.

For further information on Polity, visit our website:
politybooks.com

Contents

List of figures

Acknowledgements

We would like to thank the following for their invaluable advice and comments:

Christophe Bonneuil
Nadine Bret-Rouzaut
Michel Cassé
Pierre de Jouvancourt
Jean-Pierre Favennec
Jean-Pierre Flament
Jacques Grinevald
Kostas Kostarelos
Christine Lehman
Jean-Marc Lévy-Leblond
Baptiste Monsaingeon
Marc Monthioux
Alexandre Rojey
Jean-Noël Rouzaud
Jens Soentgen
Pierre Teissier
Sacha Tomic

Prologue
Why write a biography of carbon?

In public media as well as in climate change negotiations, carbon is in the spotlight. The talk is all about developing low-carbon or even zero-carbon levels, and 'decarbonizing' the economy and technologies alike. That verb has even made it into standard dictionaries. It seems that carbon is an enemy to be eliminated in an attempt to ensure a future for humankind. Of all the greenhouse gases responsible for global warming, it is carbon dioxide that is always being pointed at, put up for trial.

That an element as abundant, ubiquitous and familiar as carbon can become public enemy number one is not the least of the climate crisis paradoxes. Omnipresent in the media, carbon is, in fact, also omnipresent in our everyday environment. When we go back in time, we find carbon in all areas of human industry: high-tech carbons, ancient carbons, electrical carbons and even antiseptic carbons. Carbon fibres, prized for the strength and lightness they lend to the composite materials in which they are integrated, had already enjoyed industrial success with Edison's and Swan's first carbon filament light bulbs, obtained by carbonizing cotton and bamboo. Bitumen, a hydrocarbon containing 80% carbon formed from plankton slowly accumulating in sedimentary basins, was already used by the ancient Egyptians to dress roads, ships, canals, dams and reservoirs. Before the streets of London and Paris were sealed in around 1820, asphalt – a mixture of bitumen and aggregates – was used as a binder for the pigments of painters and engravers. Charcoal and bitumen have antiseptic properties that were well known to the ancients: the Phoenicians carbonized water barrels on commercial ships to preserve drinking water during their long sea voyages; the Egyptians treated dental cavities with a mixture of bitumen and clay. The alliance between humans and carbon has thus been going on for a long time, well before the industrial revolution.

Carbon's curriculum vitae is prestigious. Its work and achievements range from stellar energy to terrestrial biology going by way of the chemical industry. Born in the stars, carbon catalyses the nuclear fusion reactions by which the sun converts hydrogen into helium.[1] It is carbon, in fact, that makes the sun shine. The fourth most abundant element in the universe after hydrogen, helium and oxygen, it is also the second most abundant element, by mass, in the human body after oxygen. With carbon 'our body ... reaches as far as the stars'[2] and from the stars to our bodies, because this element, which was present at the origin of our solar system, is also the chemical basis – the molecular skeleton – of all known life. Besides the common history we share with this familiar substance, carbon is also one of the most studied elements. Indeed, carbon chemistry has a long history, with applications ranging from jewellery to heating, going via metallurgy, textiles, pharmaceuticals, electronics and green technologies. There are currently more than ten million different carbon compounds, known or synthesized.

Carbon is a record-breaking winner of prizes. The Nobel Prizes in Chemistry awarded successively to Victor Grignard (1912), Otto Diels and Kurt Alder (1950) and Herbert Brown and Georg Wittig (1979) each in their own way celebrate carbon as a genius among chemicals. Today, these names are attached to the routine reactions that make up the basic toolbox of the organic chemist. The 1960 Nobel Prize in Chemistry awarded to Willard F. Libby honours carbon-14 used in archaeological dating. The 1996 Nobel Prize in Chemistry conferred on Robert F. Curl Jr, Sir Harold W. Kroto and Richard E. Smalley is in recognition of the discovery of fullerenes. A Norwegian prize, the Kavli in nanoscience, was awarded to Sumio Iijima's work on nanotubes in 2008. In 2010, the Nobel Prize in Physics awarded to Andre Geim and Konstantin Novoselov celebrated graphene.

A fine list of achievements indeed. But is presenting the curriculum vitae of a chemical element anything more than a metaphor? It is certainly common to personify chemical elements for educational or communication purposes. Some science writers happily use this rhetorical device to enliven the periodic table. For example, in one of his bestsellers Sam Kean presents the periodic table as a storybook featuring characters who are in alliance or at war with each other. Kean emphasizes the contrast between the aggressiveness of oxygen, which bosses the other atoms, and the more friendly carbon, which has few concerns about forming bonds because it lacks four electrons to fill its outer layer and satisfy the octet rule.[3] But our aim is not to popularize chemistry. We certainly want to tell stories about carbon, but we do not need to use the atomic structure of the elements to explain their behaviour in full.

Therefore, given that we want to use an appropriate narrative style for this biography, wouldn't it be better to call a book centred on a single object a 'monograph'? For while it is true that carbon is one of the building blocks of all living things, it is not alive and does not, strictly speaking, have a life of its own. Is not speaking of a 'biography of carbon' ignoring the limits of the living and the non-living? Worse still, does it not flout the Aristotelian distinction between two meanings for the word 'life': *zoë* (the general phenomenon of life characterized by growth and decay, life and death) and *bios* (the moral and political lives of individuals, the *vita activa*)?[4]

Transgressing these boundaries between biological and political existence, between nature and culture, is precisely what carbon forces us to do. For this natural element is, as such, a hero in our culture. If it is true that human civilization began in prehistoric times with fire, by its very name carbon – which derives from the Latin *carbo* (ember) and the Indo-European radical *ker* (to burn) – constitutes an interchange between nature and culture, a meeting point between natural history and cultural history. From the pictograms traced with charcoal on the walls of prehistoric caves to the promises of carbon nanotechnologies for the twenty-first century, humans seem to have signed a pact with carbon. Carbon is a hybrid entity belonging to nature and culture. In response to the planetary crisis, we have to give up the rigid view of chemical elements as abstract, discrete and inert molecular entities and consider them as the furniture of the world that makes the earth hospitable for billions of living things.[5] Among all the elements on the periodic table, carbon is the most suited to broaden our view of chemical categories, and to reactivate them 'as cosmological forces, as material things, as social forms, as forces and energies, as sacred entities, as experimental devices, as cultural tropes, as everyday stories, as epistemic objects'.[6]

So is it really a question of narrating the tribulations of an anti-hero, a villain, the evil genius who, in the form of fossil energy buried underground, would have seduced and then destroyed the human race? A biography of carbon could position it as a diabolical being that has made a pact with humans ever since they mastered fire. The history of humankind would thus be the story of the domestication of carbon, culminating in the industrial revolution, with the extraction of tonnes of carbon buried in the subsoil, choosing 'fire engines' and machines powered by coal or oil to make up the cohort of our innumerable 'energy slaves'.[7] Having extracted, burned and consumed gigantic volumes of coal, oil or hydrocarbon gas, humans are desperately seeking to put the devil back in the box, to sequester the carbon spilled into the atmosphere in order to breathe again.[8] This grand narrative, full of mythical allusions – from Prometheus stealing fire to Faust's pact with the devil – could serve as a moral lesson to denounce the hubris of technology and encourage the development of so-called 'clean energy'.

But an edifying story for good little children does not do justice to the many faces of carbon. For carbon is polymorphic. In its elementary state alone (carbon and nothing else), it is capable of bonding with itself in multiple ways, adopting a structure that is sometimes crystalline (graphite, diamond, lonsdaleite), sometimes amorphous or nanostructured (vitreous carbon, carbon black, nanofoam) and sometimes even heterogeneous with varying degrees of order and disorder (charcoal, soot, coke). All these bodies are *allotropes* of carbon, from the Greek *allos*, 'other', and *tropos*, 'manner', multiple *ways of being*. Thus, from a chemical point of view, diamond and graphite are made up of identical carbon atoms. The only difference is the way they are bound.[9] So, here carbon is the author of two bodies with quite contrasting properties and behaviours. Diamond is hard and translucent; graphite is brittle, fragile and opaque. Diamond is abrasive; graphite lubricating. Their optical and electronic properties are very different. Yet these diametrically opposed properties are all *signatures* of carbon. It's a strange kind of thing.

Moreover, the carbon now designated as the villain poisoning our atmosphere also offers itself as a remedy. Think of the cleansing properties of activated charcoal. It's a case of carbon versus carbon. But this element is not just a two-faced Janus, because several types of carbon are being mobilized as treatments for the very ills caused by carbon. Think of all the promises associated with carbon nanotubes, those marvellous allotropic forms of carbon discovered in 1991 that are a hundred times stronger than steel, weigh six times less and can withstand high temperatures. Think of graphene, the individual sheet of carbon atoms isolated in 2004, which has extraordinary properties of strength as well as electrical conductivity. These materials of the future are supposed to provide solutions to all our problems: electronically fast; flexible; low on material cost; medicines that can be aimed at their targets like missiles; renewable energies. In short, carbon in its nano state should help us to decarbonize the economy.

Paradoxically, one of the other main tools for fighting the climate crisis is provided by carbon. It is indeed carbon dioxide (CO_2) that serves as the standard of measurement and comparison of all gases contributing to the greenhouse effect.[10] In order to assess this contribution quantitatively, each of these gases is assigned an index, its 'global warming potential' (GWP). This index is based on the CO_2 = 1 unit of measurement, over a century. Carbon therefore plays a key role as a commensuration tool: it provides a common measure, it offers a handle on quantification. It allows us both to compare greenhouse gases with each other and to compare *our actions* with each other: flying, planting a tree, buying a prime steak, sending an email, attending a football match. All of these activities can be equated to establish carbon balances and enter into a global and disproportionate accounting, fitted to the scale of the earth system.[11] The so-called 'carbon equivalent' is now the main currency of exchange between human industry and the planet in a trade that includes financial transactions such as carbon offsetting and carbon trading. Carbon is thus becoming a general equivalent and a tool for calculation and action to regulate our exchanges with the environment.

The alliances that human societies have made with carbon are multiple, and are not limited to the 'choice of fire'.[12] From the dawn of humankind, carbon's destiny seems to be linked to *writing*, to its modes of inscription, fixation, standardization and circulation. From the charcoal used for the drawings inscribed on the walls of caves to the graphite pencil (from *graphein*, 'to write') and then on to graphene electronic chips, from diamond points as engraving tools to the 'code of life' based on a carbon skeleton, from the periodic table of elements to carbon accounting, from Otzi the Iceman's tattoos made of black smoke (5,200 years ago) to the global warming potential measured in carbon equivalents, via carbon-14 dating, everything is happening as if carbon offered humans a way of inscribing their history into the temporalities of nature, technology and the cosmos that are infinitely smaller, or bigger, than their own.

5

It is a case, then, of conjugating all these temporalities, mixing these heterogeneous histories by deploying some of carbon's figurations and adventures. So, this book is not a biography in the sense of a chronological account, from the birth to the death of an individual. It could certainly recount the life trajectory of a carbon atom from its birth in the stars to the present day (without anticipating its death). This is the kind of story Primo Levi sketches out in the last chapter of his book *The Periodic Table*, a genre taken up by several science writers.[13] Carbon lends itself to these atom-centred narratives: 'I, a carbon atom, on my travels around the world . . . this happens to me, then that, etc.' This kind of biography is probably very effective in highlighting the inclusion of human history in the dynamics of the earth system, but it presupposes that the identity of carbon can be summed up by carbon atoms. The nature of carbon is taken as given, determined by the structure of its atoms, and these entities are then engaged in narrative excursions.

This book offers a completely different perspective on carbon. Far from presupposing from the outset that it is a natural substance well defined by the nature and structure of its atoms, it begins by asking how this being came into existence. How did we come to gather under the same element things as diverse as the nauseating emanations known since antiquity as 'mephitic air',[14] black coal and the diamond crystal, symbol of purity and durability? How did we come to know the structure of carbon atoms and understand how they are bound? The first question is how carbon came to be the chemical element that all science books discuss, how it inspired a chemistry all of its own and how it still continues to hold the kinds of surprises that feed Nobel prizes. But it is also a question of *in what way it exists as* an element, a material, a gas, a technical device, a currency of exchange or a pencil mark.

For this biography does not prioritize the chemical definition of carbon. It postulates that the *vita activa* of carbon is also understood and described in detail through its participation in the rise of civilization, the industrial revolution and consumer society. The aim is to specify not *what it is* (its nature) but *in what way it is* (its *modes of existence*).[15] The aim is also to show the diversity of carbon's modes of existence – chemical, geological, biological, cultural, technical, economic, geopolitical, etc. – without presupposing a fundamental ontological stratum underlying this range of manifestations. In short, this biography is a way of experimenting with a new style of metaphysics in the case of carbon. It invites us to swap an older ontology for an *ontography*. Unlike ontology, the general science of being, which aims to order the entities that make up the universe in the clutches of a grand theory, ontography is more concerned with putting the ways of being of singular things into narrative form. It unfolds the range of relationships that these things weave around and among themselves, as well as the modes of inscription that they offer to our material and symbolic practices.

This book wagers that narrative style prevails over argumentative discourse when it is a matter not of defending a thesis, but of opening our eyes to the world we live in so as to unravel the problems by going beyond grand discourses or catch-all formulas such as 'decarbonization'. So let's make room for stories and histories.

PART I

THE INVENTION
OF CARBON

1

Mephitis

At the foot of high mountains, in the middle of Italy, there is a well-known place, whose fame has spread to many lands, the valley of the Amsanctus. A dark forest presses in upon it from both sides with its dense foliage and in the middle a crashing torrent roars over the rocks, whipping up crests of foam. Here they point to a fearful cave which is a vent for the breath of Dis, the cruel god of the underworld. Into this cave bursts Acheron and here a vast whirlpool opens its pestilential jaws, and here the loathsome Fury[1] disappeared, lightening heaven and earth by her absence.[2]

Thus, in the *Aeneid,* Virgil describes the complex of springs and fumaroles in the Ansanto valley (*Amsanctus* in ancient times), near the Hirpine mountains in Campania, central Italy (Figure 1). There one finds the small, constantly boiling lake (*laghetto dei soffioni*), a gorge, the Caccavo river and a ravine into which the Caccavo pours its waters, the *Vado mortale.* The Samnites, an Oscan-speaking tribe established in central Italy from the sixth to the second century BCE, are said to have built a temple there dedicated to a pre-Roman goddess of emanations, volcanic springs, caverns and fumaroles: Mephitis.

The name has remained in the designation of the place, the *Mefite di Rocca San Felice.* By metonymy, we also speak of 'mofettes' as another name for fumaroles. The adjective 'mephitic' was also used to describe the 'air', the 'spirit' or, later, the 'gas' that we refer to today as 'carbon dioxide'. It has also been called a 'lethal' or 'deadly' spirit, a 'sylvan' or 'wild' spirit, as well as 'fixed air', 'dephlogisticated' or 'rotten'.

The Mefite is still as active as ever. This is one of the most constant and abundant natural sources of carbon dioxide on earth. Recent geophysical studies show that the waters from the Mefite lake come from several kilometres underground, a carbonate substratum rich in compressed gas pockets released during tectonic events dating from the Messinian (between seven and five million years ago).[3] This particular pocket emits

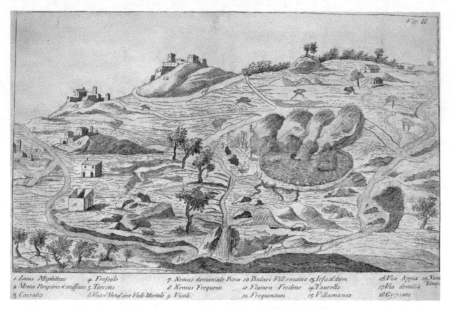

Figure 1. The *Lacus Mephiticus* and its surroundings. Vincenzo Maria Santoli, *Roccae Sancti Felicis. De Mephiti et Vallibus Anxanti, Libri Tres*, Naples, 1783.

on average two thousand tonnes of gas per day, including carbon dioxide (CO_2), but also methane (CH_4) and hydrogen sulphide (H_2S). It is the sulphur in the latter – and not the carbon – which gives the Mefite its characteristic rotten egg smell. The first two carbon gases are invisible and odourless, but they are just as active.

Virgil's description highlights the spectacular and frightening character of the place, and for good reason. The site is actually very dangerous, especially when there is no wind. Nowadays, there is a sign to warn the curious: *Pericolo di Morte* (Danger of Death). There are often corpses of birds and small mammals that have accidentally come to drink. Many incidents with humans have been reported over the centuries. In the 1990s, three people died there. Today, the Mefite is sometimes treated as a natural analogue, a life-size model of 'leaking CO_2',[4] useful for the study of leakage risks in future geological storage sites.

A thing with many names

Is it acceptable to consider Mephitis to be an ancient personification of our modern carbon dioxide?[5] And does this imply that only the *names* change while the thing remains the same?

The two beings – the goddess and the chemical substance – certainly meet very different identification criteria: on the one hand, an indomitable supernatural power; on the other, a well-defined and characterized chemical substance. To consider Mephitis as a name used by the Romans to 'designate' carbon dioxide would simply be anachronistic. Firstly, the term 'gas' did not exist then. The Latins used *spiritus*, translating the Greek πνεῦμα (*pneuma*): exhalation, breath, respiration.[6] Secondly, they did not distinguish, among these 'breaths', those which are composed of carbon and those which are composed of sulphur (the latter being precisely responsible for the odour described as 'mephitic'). Finally, although the Latin *carbo* is attested as meaning 'ember' or 'coal', the ancients did not seem to associate it with Mephitis' breath.

It would also be confusing different modes of spatial inscription and thus be guilty of 'anatopism'. On the one hand, the *genius loci* associated with singular places; on the other, a mobile, deterritorialized, globalized, even universalized entity – in a word, modern. Mephitis is associated with specific local *climates*, but CO_2 with the climate itself.

It would therefore be both anachronistic and anatopic to assert that the ancients 'metaphorically' represented, in the guise of a goddess, what was 'really' carbon dioxide. This would ignore the very real phenomena that took place in the places associated with the goddess while suggesting that our modern CO_2 – that evil genie out of the box – would be purged of all metaphorical connotations. No, Mephitis did not emerge in the same time or space as carbon dioxide. The two terms are not interchangeable.[7]

And yet, reading the stories about Mephitis as if they were talking about carbon dioxide seems to us to constitute a narrative move that is meaningful for our investigation into carbon's modes of existence. But we need to specify why.

We want to bring the two terms together by inscribing this thing in a world where earth and air are exchanged. Mephitis is not a name that designates or represents a particular gas. It is a way of telling the story of a powerful action, a fearsome breath that makes a place inhospitable, uninhabitable. It is a *signature* through which the gas is inscribed in the world, distinguishing different *heterotopic* places. It is therefore less a question of seeing Mephitis as the source or origin of the modern concept of CO_2 than of recounting how, by giving *realms of memory a signature*, the breath of Mephitis has been inscribed in popular language and culture, and then reinscribed in the operating knowledge of chemists and even in the current climate crisis.

A genius of place

What kind of goddess was Mephitis? She is not considered a major deity in the Latin pantheon, perhaps because of her barbarian (non-Greek) origins. According to Virgil, the Romans included her in the nebula of the Erinyes or Furies, the female deities responsible for persecuting the victims of fate, and therefore essentially evil. But the Latin authors themselves only report bits and pieces of anonymous stories, borrowed from the peoples who lived there before them. Thus, although the Ansanto Mefite is referred to not only in Virgil and his commentator Servius, but also in Tacitus, Seneca, Cicero, Ovid, Tertullian, Horace and Pliny the Elder, none of these texts can be identified as 'the' source.

The etymology of the name 'Mephitis' is controversial. Some experts claim that it has to do with intoxication: Mephitis is said to be 'the intoxicating one'. For others, it means 'the one with smoke in the middle', or 'who stands in the middle', or even 'who holds the middle' (from the Oscan *mefiu*, median). According to the palaeolinguist Michel Lejeune, her name is based on *fonti*, 'source', or 'fountain': Mephitis would thus be the 'goddess of sources'.[8] Archaeologists now believe that her cult was widespread throughout central Italy from Pompeii to Rome, that it was linked to the volcanic nature of the region and that its centre was the Ansanto site.[9] Thus Servius, in his commentary on Virgil's *Aeneid*, goes so far as to assert that:

> Topographers call this place 'the navel of Italy'. It is indeed on the border of Campania and Apulia, where the Hirpini are; it has sulphurous waters, and worse smelling for the reason that it is surrounded by forests. An entrance to the underworld is said to be there, because the terrible stench kills those who approach it, to the point that victims were not sacrificed in this place,

but when they were put into the water the stench caused them to perish. This was a kind of sacrifice.[10]

Although the memory of Mephitis is lost in the mists of time, traces of her have never faded.[11] Maps from the seventeenth and eighteenth centuries even indicate and locate the temple of Mephitis quite precisely.[12] When the 'new chemistry' of the eighteenth century took off in Naples, its main laboratory was Vesuvius, so much so that Neapolitan chemistry was first and foremost a Mephitian chemistry, a volcano chemistry rather than a laboratory one.[13] Then, as now, the vitality of the memory of Mephitis has less to do with the perpetuation of archaic superstitions than with the volcanic nature of the places and the 'breaths' that manifest themselves there. For the stories of the ancients are all topo-narratives: they make *places* speak. Carbon dioxide thus configures these places. It is always associated with localities, and in cultural memory these places are linked with death.

Why is this gas associated with death? Carbon dioxide is not toxic in itself – otherwise drinking soda water would be dangerous – but at a certain concentration it makes people dizzy, and then at high doses fatally asphyxiates them: hence the 'lethal spirit' name in some texts. However, certain natural phenomena can make CO_2 an invisible and silent mass killer. Such as limnic eruptions: the sudden degassing of a deep lake, often of volcanic origin, whose lower and upper waters barely mix and form layers strongly differentiated by their concentration in compressed gases. In 1986, the collapse of a rock face released a layer of carbon dioxide that had been stagnating for centuries at the bottom of Lake Nyos in Cameroon. This outgassing, equivalent to the explosion of a one cubic kilometre CO_2 bubble, instantly decimated more than 1,700 people, thousands of livestock and countless wild animals, all killed by asphyxiation in an area of several kilometres.[14] The cataclysm was of such magnitude and strangeness that conspiracy theories abounded (some imagined the explosion of a neutron bomb). However, limnic eruptions are not uncommon in the volcanic regions of Central Africa and are often referred to by the Swahili term *makuzu* (evil wind). For some local people, the *makuzu* of Lake Nyos was a revenge of the spirit of the lake against villagers who were disrespectful towards it, as portrayed by the Cameroonian playwright Bole Butake in *Lake God.*[15]

For others, it was the work of Mami Wata, the powerful water goddess. Is there an African Mephitis? There are local spirits[16] in all civilizations (except ours, it seems) and some are closely associated with the thunderous and sometimes deadly eruptions of underground gaseous carbon compounds.

Geomythologies

Even among authors considered more 'naturalist' than Virgil, the crafter of legends, natural phenomena are always qualitatively different in different places. This localization gives them a flavour of myth, of geomythology.[17] In his *Natural Histories*, Pliny devoted a chapter to the 'products of the earth … emanations'. He used the same term as Virgil, *spiracula*: 'places called breathing holes [*spiracula*] or by other people jaws of hell [*Charonea*]';[18] 'ditches that exhale a deadly breath; also the place near the Temple of Mephitis at Ampsanctus in the Hirpinian district'. Pliny also mentioned 'the hole at Hierapolis in Asia', a spa in Asia Minor built on a geological fault, 'harmless only to the priest of the Great Mother'.[19] As for Cicero, he mentioned both the 'mortal lands' of *Amsanctus* and those of the *Plutonia* of Hierapolis. In his treatise *On Divination*, he associated the qualitative difference among lands and their climatic effects on human behaviour with the breath of the gods, sometimes spread in the subsoil, sometimes in 'inspired' individualities:

> Do we wait for the immortal gods to converse with us in the forum, on the street, and in our homes? While they do not, of course, present themselves in person, they do diffuse their power far and wide – sometimes enclosing it in caverns of the earth and sometimes imparting it to human beings. The Pythian priestess at Delphi was inspired by the power of the earth and the Sibyl by that of nature. Why need you marvel at this? Do we not see how the soils of the earth vary in kind? Some are deadly, like that about Lake Ampsanctus in the country of the Hirpini and that of *Plutonia* in Asia, both of which I have seen. Even in the same neighbourhood, some parts are salubrious and some are not; some produce men of keen wit, others produce fools. These diverse effects are all the result of differences in climate and differences in the earth's exhalations.[20]

Cicero suggested that while the Pythia utters her oracles under the influence of Apollo's fumes emanating from the ground, the Sibyl performs her dignified function as a prophetess in full possession of her faculties. But in the *Aeneid* Virgil depicted an ecstatic Sibyl, a woman 'in wild frenzy'.[21] It should be remembered that Aeneas' dead father, Anchises, had appeared to his son and urged him to consult the Sibyl; she was to guide Aeneas to the entrance of the underworld so that Anchises could show him his future descendants. In his funeral appearance, Anchises locates the entrance to the underworld in Lake Averno, an ancient crater lake in the Phlegrean Fields (literally 'burning fields'), a volcanic area extending to the west of Naples. It is in nearby Cumae that Aeneas came to consult the Sibyl, in a temple of Apollo that Virgil describes more accurately as a 'cave' or 'den'. When 'the god came to her in his power and breathed upon her', her breast heaved and her face transfigured.[22] This is, on the evidence, the effect of gases going to her head.

Some geologists have linked Pliny's[23] or Cicero's[24] accounts of the ecstatic behaviour of the Pythia of Delphi with an ancient leak of carbon dioxide and hydrogen sulphide into a seismic fault.[25] Whether one is with Pluto in Ansanto or Hierapolis, or with Apollo in Delphi or Cumae, Mephitis is not far away, and there is CO_2 in the air.

Mephitis is also known to lodge in caves. Poisoning by 'heavy air' (carbon dioxide) has been one of the dangers feared by miners and cavers alike since ancient times. Caves, especially limestone caves, are generally etched out by acidic water because it is rich in carbon dioxide. A large quantity of this gas can remain trapped in a cave or reach other caves through cracks. This is the case of the *Grotta del Cane*, located in the Phlegrean Fields, a few steps from Lake Agnano (drained in 1870), whose waters were constantly boiling, and not far from the Solfatare ('land of sulphur'), an ancient dried-up crater lake. The cave is named after the custom of using a dog to show if mephitic exhalations were present. Dogs in distress were taken from the cave and thrown into the Lake of Agnano and they came out fully revived (Figure 2).

At the height of the Enlightenment, Diderot and d'Alembert's *Encyclopédie* devoted an article to the Dog Cave, recounting how many animals and humans were exposed to the continuous stream of its deadly exhalations and with what effects, which were very precisely described.

Le Lac d'Agnano.
1.Sudat.S.Germano. 2.Grotta de l'Cane.

Figure 2. Lake Agnano and the *Grotta del Cane*.

They make the connection with other similar places – mines, caves – in Hungary, Sicily and Italy.[26]

Finally, earthquakes can also create faults that release compressed air pockets from the subsoil to the surface. This is what Seneca wrote in Book VI of his *Natural Questions*, dedicated to 'earthquakes'. As a good Stoic, Seneca explained earthquakes by material causes: the flatulence of the earth, which expels the overflow of winds that it has absorbed and retained in its entrails. 'The earth is naturally porous and has many voids. Air passes through these openings. When air in large quantities flows in and is not emitted it causes the earth to tilt.'[27] Even though it is often translated as 'air' or 'wind', what Seneca calls *spiritus* is not entirely confused with simple air, that which we breathe (*aer*). While being material, this spirit-air is perceived above all by its effects. It is, for Seneca, the most powerful and energetic element of nature, without which the others would have no strength. Without it, there is no fire; without it, water sleeps. It is the element that reeks of earth. Its dynamic characteristics explain the curious phenomena that accompany earthquakes, such as the unexplained death of six hundred sheep in Pompeii the day after an earthquake.[28]

Compared to Virgil's text, Seneca's very naturalistic explanation seems to banish mythology by erasing any reference to the 'hated goddess'. The tone is medical, even hygienic. From mephitic breaths, Seneca moves on to 'pestilential vapour' and 'noxious air':

In several places in Italy a pestilential vapour is exhaled through certain openings, which is safe neither for people nor for wild animals to breathe. Also, if birds encounter the vapour before it is softened by better weather they fall in mid-flight and their bodies are livid and their throats swollen as though they had been violently strangled.

As long as this air keeps itself inside the earth, flowing out only from a narrow opening, it has power only to kill the creatures which look down into it and voluntarily enter it. When it has been hidden for ages in the dismal darkness underground it grows into poison, becomes more deadly by the very delay, becomes worse the more sluggish it is. When it has found a way out it lets fly the eternal evil and infernal night of gloomy cold and stains darkly the atmosphere of our region. For the better is conquered by the worse. Then even pure air changes into noxious air; from it come sudden and continuous deaths and monstrous types of disease arising from unknown causes. Moreover, the disaster is brief or long according to the strength of the poison, nor does the pestilence stop before the clearing of the sky and the tossing of the winds have purified that deadly air.[29]

Yet the emphasis on natural causes does not change the connotations associated with mephitic breaths: death, darkness, plague, calamity, stench, poison, etc. Thus, from mythical history (Virgil) to natural history (Seneca), the accounts of the ancients seem to seal from the outset the identification of carbon dioxide with a demonic being escaping from the infernal depths.

From late antiquity, the theonym Mephitis gave way to the adjective 'mephitic', which designates anything that is stinky, fetid, unhealthy and nauseating: from the skunk, a cousin of the polecat belonging to the *Mephitidae* family, to the famous rue Mouffetard in Paris, whose name comes from the stinking waters (*mouffeteuses*) that ran down its slope. The French slang word *moufter* is a catch-all verb meaning to denounce, protest, sniff and smell bad. In Christian writers such as Augustine, the toponym *Amsanctus* also became a synonym for 'bad smell', even though – ironically enough – it is etymologically close to 'sacred' and 'sanctuary'.[30] Christians made Mephitis one of the names of the devil, from Dante's Mephisto to Goethe's Mephistopheles.

But why worship a 'hated goddess' to the point of building temples to her? Is Mephitis really to be feared?

An elixir of youth

Not everything that is *mephitian* is *mephitic*. The focus of 'mephito-clasts' of all kinds on smoke and on everything, as in Seneca, to do with elemental air tainted by the earth conceals the no less important association of Mephitis with elemental water. And what if the Romans had already demonized the Oscan goddess before the Christians did?

The waters of Mephitis are not all 'mephitic', far from it. For example, another Mephitis sanctuary unearthed at Rossano di Vaglio (Lucania) was built near a pure water spring and even *for* this spring, whose waters crossed the main piazza in an open-air pipe. Mephitis was also associated with the beneficial virtues of thermal waters before the Madonnas of the baths in the Naples area took over with the expansion of Christianity.[31] Called both 'god Mephitis' and 'goddess Mephito', she also embodied vegetarianism and androgyny: the priests who venerated her did not eat meat and dressed in female clothes. Some even identified Mephitis with the Venus of Pompeii,[32] associated with clear fountain water and rejuvenating baths. For Lejeune, the inscriptions found in the Lucanian sanctuary of Rossano plead in favour of a beneficent Mephitis, even a healer. She watched over peasant life, the fertility of women and the fertility of the fields, and protected ploughing and animals. Even long after the Mephitis of Ansanto cult ended and the sulphur mine was installed under Napoleon's reign, shepherds, despite the recognized danger of the site, continued to take their sick sheep there during the transhumance.

The Pluto–Apollo duality noted above – dark god of the depths on the one hand, celestial god of divination on the other – is a good reason for Mephitis' essential ambivalence. Source or breath, healer or poisoner, Mephitis is a mediator between the underworld and the upper world. She makes the two realms communicate, ensuring the passage from one to the other. She 'stands in the middle' between several kingdoms:

between fertility and death, masculine and feminine, savage and civilized, the elemental pairs earth–air and water–fire, the Chthonic depths and the Uranian air. As Virgil writes, she helps the earthly to gain solace in the heavenly. Conversely, she absorbs their wrath and fertilizes the land. Mephitis is a *pharmakon*: both poison and remedy, poison *because it is a* remedy, remedy *because it is* poison . . . and also a scapegoat – like CO_2 is for us today. We are therefore tempted to choose, from among the hypotheses on the etymology of Mephitis, the one that presents it as 'standing in the middle' between myth and science. Mephitis is a fine example of myth and science intermingling and shows how the two can nourish each other.

2

An indescribable air

From mephitic air to sylvester spirit

If you burn 62 pounds of charcoal, you get 1 pound of ash. This means that 61 pounds of *spiritus silvestris* (wood spirit) has been released. Thus was the claim made by Jan Baptist Van Helmont (1579–1644). This Flemish physician and chemist introduced the word 'gas' to denote that 'something' which is produced in combustion but which cannot be seen or contained. Thus the spirit of wood is the archetype of gas.

It seems that the scenery has changed completely. We have left smouldering caves for the laboratory, the world of myth for experimental science. The term 'gas' is introduced in the context of a critique of the theory of the four elements, rejected in favour of a theoretical view based on the states of matter (solid, liquid and, of course, gaseous), that clearly mobilizes the principle of the conservation of mass in chemical reactions.[1] Van Helmont's view sounds so modern that one could think that the introduction of the term 'gas' chased pestilential vapours away and marked the dawn of positive science.

But this is not the case. This episode forces us to revise the science-as-enemy-of-myth fable. It is true that Van Helmont, unlike former alchemists, claimed that the transformation of water into steam was not a transmutation, but rather a division into *minima partes* (minimal parts). Certainly, he got a glimpse of the notion of states of matter when he emphasized the distinction between physical changes and chemical transformation.[2] But the word 'gas' does not dispel the mystery of the harmful powers associated with mephitic air. According to the etymologies, this neologism is inspired either by the High German *Geist* (spirit) or by the Greek χάος (*khaos*: literally 'fault', 'gap'), a concept associated by the first Greek physicists with the indistinct state (called 'inchoate') of the elements that preceded the organization of the world. However, Van Helmont uses it not to designate the gaseous state *in general*, whatever the substance, but to qualify the spirit of wood and *it alone*. He thus differentiates it from the ambient air and water vapour released during the processes of combustion and putrefaction. The spirit of the forest is a wild spirit (*silva, selva,* sylvester [forest] and *sauvage* [wild] have the same etymological root) that is invisible and intransigent, always escaping us and sometimes exploding. As it emanates from the processes of both life (breathing) and death (putrefaction, combustion), it is considered a vital *and* lethal principle, anabolic and catabolic. An internal principle of animation, powerful, subtle and elusive. Always in motion, it proves to be extremely dangerous once extirpated from the bodies that it animates and that contain it.

Wood or sylvester gas thus retained Mephitis' features, although it was mobile, relocatable, circulatory, no longer having to reside in key places such as the Ansanto Mefite, Pythia's lair or the Dog Cave (mentioned by Van Helmont). This Flemish Mephitis no longer lived in caves near volcanoes. It spread out over wetlands and woods, because it was present in every living organism, and could be released through combustion and fermentation.

It is in this sense that seventeenth-century physicians, confronted with the plague of London in 1665, appropriated the concepts of gas and wood spirit indiscriminately. This gas was the material cause of the plague, according to the pharmacist William Boghurst:

> [T]he matter of the plague is no other than *fracidum gas terrae* [putrefied earth gas] as he [Van Helmont] calls it; that is, as he himself explains, *silvester spiritus veneno tinctus* [a sylvan spirit tinged with venom] or a poisonous vapour which rises from the earth, and this is the opinion of many learned physicians.[3]

Clearly, the invention of the word 'gas' does not displace the constellation of meanings and values surrounding the original identity of mephitic air. It suggests, however, that this thing of various names (mephitic air, lethal spirit, wood spirit or sylvester spirit) was of interest above all to doctors. These 'chymists' were more concerned with health than with the structure or the states of matter.

From sylvester spirit to fixed air

This reputedly evil air invented a new identity for itself via the manipulations of two eighteenth-century physicians. Stephen Hales (1677–1761) initiated the science of gases – pneumatic science – through his experiments on plants. This air, which was once associated with specific localities, he found in the leaves, branches and trunks of trees. He claimed that it was contained in plants since by burning them he could release it and collect it by means of an ingenious apparatus that he developed. It consisted of a glass bell jar inverted on a liquid and connected by a pipe at the base of a flask where a chemical reaction was taking place (Figure 3).

Hales called this air imprisoned in plants and solids 'fixed air' to distinguish it from atmospheric air, which experimental philosophers, such as Robert Boyle in particular, had described as elastic and compressible. While stressing that air is Protean and ubiquitous, Hales characterized its

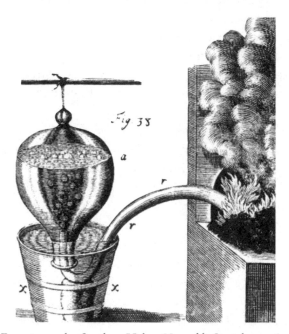

Figure 3. Experiment by Stephen Hales, *Vegetable Staticks, or, An Account of Some Statical Experiments on the Sap in Vegetables ...*, London: Printed for W. and J. Innys; and T. Woodward, 1727.

diversity in exclusively physical terms, and he invited philosophers to ask themselves whether air might not also be an eminently active chemical principle.

The message seems to have been heard because pneumatic science provoked a craze among natural philosophers in the last decades of the Enlightenment, to the point that historians describe this episode as a 'hunt for airs'.[4] This hunt began with a medical student, Joseph Black (1728–99), who studied in Scotland. The universities of Glasgow and Edinburgh had well-equipped laboratories. In the 1750s, Black undertook research under the direction of William Cullen for his thesis on *magnesia alba* (magnesium carbonate, $MgCO_3$). He found that this substance, used as a purgative by physicians, behaved like limestone. It effervesced under the action of an acid, thus releasing the fixed air it contained; it also released fixed air when calcined, thus producing quicklime (calcium oxide).[5] But Black noted that the lime resulting from the calcination of *magnesia alba* was not caustic, like the limes of limestone. He concluded that the causticity of quicklime was not, as previously thought, the product of calcination, but came from an intrinsic property of limestone that *magnesia alba* did not possess. However, Black was less interested in the composition of this *magnesia alba* than in the fixed air that it contained. He then undertook a series of experiments to determine the affinities of the fixed air. Here was a new player on the chemical scene: the fixed air captured in the glass bells captivated a whole generation of chemists.

Following Black, several scientists continued to investigate the behaviour of fixed air: David McBride showed that it inhibited the breathing of animals; Henry Cavendish studied its combinations with several substances; and Joseph Priestley demonstrated that cold water could absorb a volume equal to its own. He invented soda water. He also established that this water impregnated with fixed air had a sour taste and that it turned blue litmus paper red. Far from being evil, it would have beneficial effects on health, as confirmed by some natural mineral waters. This was Priestley's selling point when he marketed Pyrmont water (the first artificial sparkling water) for the British navy.[6] The fixed air purified the water of germs and kept it fresh and healthy during long sea voyages. It was even supposed to protect sailors from scurvy.

But if fixed air was an unbreathable air, stale, spoiled, even 'rotten' because its phlogiston was evacuated, what was the source of its revitalizing power? Priestley argued that it came from plants. If fixed air was degraded air, could it not be restored to allow respiration again? Priestley's experiments on plant respiration were intended to address this hope, which was motivated by health concerns. '[O]n the 17th of August 1771, I put a sprig of mint into a quantity of air, in which a wax candle had burned out, and found that, on the 27th of the same month, another candle burned perfectly well in it.'[7] Of course, Priestley had chosen mint for its refreshing properties, but he extended the experiment to other, less pleasant-smelling plants, with the same result. He concluded that plants were able to restore fixed air, to 're-phlogisticate' it to make it breathable again. The discovery of this exchange between the animal and the vegetable kingdom represented, for Priestley, a dazzling confirmation of the natural religion that he was promoting. It demonstrated the remarkable harmony that reigned in the economics of nature. The President of the Royal Society, Sir John Pringle, commented on Priestley's findings as follows:

> From these discoveries we are assured, that no vegetable grows in vain, but that from the oak of the forest to the grass of the field, every individual plant is serviceable to mankind if not always distinguished by some private virtue, yet making a part of the whole which cleanses and purifies our atmosphere. In this the fragrant rose and deadly nightshade co-operate: nor is the herbage, nor the woods that flourish in the most remote and unpeopled regions, unprofitable to us, nor we to them considering how constantly the winds convey to them our vitiated air, for our relief, and for their nourishment.[8]

In just a few years, fixed air established its own elaborate profile, distinct from other airs, with a brand image popularized by the benefits of natural gaseous waters and pride of place in the economics of nature, a currency of exchange between animals and plants: fixed air produced by animals came to nourish plants that animals needed to eat, and plants revitalized the air so that they could breathe deeply.

Fixed air was thus at the top of a long list of air fluids isolated, and then identified, during the decades to follow: flammable air (future hydrogen), vital or dephlogisticated air (future oxygen), phlogisticated air (future nitrogen).

From fixed air to carbonic acid

The hunt for airs kept scientists busy all over Europe and ended up killing off phlogiston, the key element contained in combustible substances as well as in metals; it fell victim to assaults from Antoine Lavoisier in 1775. This episode, known as 'the chemical revolution', has been recounted many times and focuses on the phenomenon of combustion, which Lavoisier approached with the help of scales, carefully weighing everything going in or out of the reaction chamber; in short, by doing the maths.[9] Less well known is the important role of fixed air in this affair.

The 'summary historical account of the elastic vapours' that opens Lavoisier's 1776 *Essays Physical and Chemical* clearly shows that he was following in the footsteps of Black's and Priestley's work on fixed air, which had been introduced into France by Guillaume-François Rouelle, Jean Bucquet and Antoine Baumé. Lavoisier's own early research, recorded in these essays, deals with fixed air, its link with quicklime and with wine fermentation. He suspected that it might also have a link with what 'passes from the earth in vegetation, by means of that motion of universal fermentation which the return of the sun excites in nature at the beginning of spring'.[10]

More tortuous was the route that led Lavoisier from the study of the properties of fixed air to the determination of its composition. While repeating the gas experiments of his colleagues from various countries, he tried to establish the nature and the proportion of the components of each reaction product, thanks to gravimetric analyses. The composition of fixed air was neither immediately nor directly determinable; it had to be interpreted via combustion and calcination, both phenomena understood as combination with part of the air and consequently as decomposition of atmospheric air. Lavoisier then went through the determination of the properties of vital air, which he considered as a principle common to all acids, and therefore named the 'oxygine [*sic*] principle' (acid generator). From then on, the fixed air resulting from the combination of this principle with the coal of fuels was redefined as 'carbonic acid'.[11]

This rough historical sketch of the identification of the gas that we call carbon dioxide today is an *a posteriori* reconstruction. More precisely, this historical approach, informed by what we know of its properties today, is a way of paying attention to how carbon dioxide has been repeatedly redefined in various times and places. We thus highlighted its multiple ways of being-in-the-world, successively as a goddess, a power attached to localities, a more or less evil spirit, an air that can be captured, measured and dissolved and an acid whose affinities can be determined. Through its various modes of existence, with different characteristics and footprints, this mephitic air points to the existence of carbon, without actually involving it.

3

Between diamond and coal

The flurry of adjectives circulating around the gas – mephitic, sylvester, fixed and then carbonaceous – is a retrospective sign of the presence and importance of carbon in science and in the natural economy of the Enlightenment. But was this the way that carbon came out into the open?

Despite the importance of the work on fixed air, we have to look elsewhere to see how carbon came into existence as a term in the course of a project to reform the language of chemists. We have to follow the trail of the diamond, which was the object of a major campaign of enquiry right after fixed air. However, we shall see that this 'pure carbon' was still impregnated with black coal towards the end of the eighteenth century.

The diamond enigma

On 26 July 1771, in front of a crowd of scientists and jewellers, Pierre-Joseph Macquer, a famous chemist and member of the Royal Academy of Sciences in Paris, placed a diamond given to him by the king's equerry, Charles-Théodore Godefroy de Villetaneuse, into a porcelain furnace from the Sèvres factory and heated it. The diamond turned bright red and then disappeared entirely: 'not the slightest vestige of it remained', noted the register of the Academy.[1] In short, diamonds can be burned.

This was not, strictly speaking, a discovery: for nearly a century in the academies of London and Florence, as in Paris at the Jardin du Roy, diamonds had been taken to high temperatures, and it was noted there were vapours, discolouring, loss of lustre and weight and then dissipation. Nevertheless, it was difficult to believe that 'the hardest of all the products of nature'[2] could be volatilized in this way. Its name, derived from the Greek ἀδάμας (*adaman*, indomitable), speaks of its resistance to erosion over time as well as to attempts to break it. This is why diamonds are a symbol of purity and durability, an alliance guaranteed to be capable of withstanding all the assaults of time as well as the vicissitudes of fortune. Isn't it said that 'diamonds are forever'?

So why were scientists obstinately trying to destroy this rare and precious stone? The strange behaviour of diamonds aroused the curiosity of Parisian society: bourgeois and nobles offered diamonds and precious stones from their cabinets of curiosities to be burned, and they swelled the crowd of spectators – geologists, chemists, jewellers – at the experiments conducted at the Royal Academy of Sciences. The nature of diamonds was a hot topic in scientific circles throughout Europe during the last two decades of the eighteenth century. It mobilized talents, a few diamonds sacrificed to science and a major instrument: the burning mirror made in Germany in 1702 by Walther von Tschirnhaus for the Duke of Orleans, then given to the Count of Onsembray, who bequeathed it to the Royal Academy of Sciences in 1752. The apparatus was made with a huge biconvex lens, 33 inches in diameter and weighing 160 pounds, and a second lens was added to shorten and concentrate the focus, and equipped with a mechanism to follow the movement of the sun (Figure 4). Spectacular experiments could be carried out with such an instrument. The challenge was to expose the diamond to the intense fire of the sun concentrated by the lenses' focal point, thus subjecting it to the test with 'pure' fire, i.e. without coal. Macquer, Lavoisier and Louis-Claude Cadet de Gassicourt set up a series of experiments to determine whether it was really volatilization or some other phenomenon.

Jewellers skilled in the art of cutting and cleaving diamonds were accustomed to subjecting them to strong heat to eliminate defects, without making them volatile. One of them, Stanislas Marie Maillard, did not hesitate to place his eight diamonds, carefully packed with coal powder in a well-fired pipe, inside Macquer's furnace. After several hours of heating, 'the eight diamonds still had their facets, their polish, in a word, they were just as they were before the operation'.[3] At the Academy's Easter 1772 public meeting, Lavoisier suggested that 'this unique substance is not really as volatile as we thought', but that, in contact with air, something like combustion occurs.[4] The presence of air suggested, in fact, that it could be combustion, a phenomenon that Lavoisier was working on with the help of scales, weighing inputs and outputs.

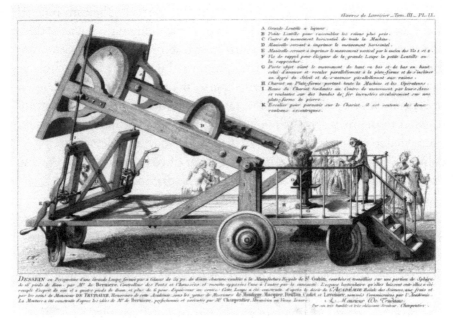

Figure 4. Burning mirror from the Royal Academy of Sciences, *Œuvres de Lavoisier*, Vol. III.

In 1773, Macquer, Lavoisier and Cadet, together with Mathurin Jacques Brisson, launched a new series of experiments to detect and then analyse any air that might be released from the combustion of diamond. They had an apparatus like the one used by Hales and called on a host of collaborators because many hands were needed to insert the substances, manipulate the lens, observe, note and dictate, under dazzling light and intense heat, the changes in colour and odour, collect the fumes and then analyse them. This was what we would call Big Science today. Was the mobilization of all these resources sufficient to understand diamond's nature?

In 1776, Lavoisier published a memorandum on diamond, presenting the results of their experiments: diamond is a combustible body; it burns in open air with a luminous flame; the volume of air decreases; and the vapours produced by diamond are something like the fixed air given off by burning coal. Moreover, when the heat is not too intense and the air is not renewed, 'the diamond is reduced to a black and carbonaceous matter'. Lavoisier believed that the carbonaceous matter covering the diamond came from impurities or surrounding bodies.[5]

In their perplexity, the academicians blamed the tool. They announced that they needed a more powerful lens to pierce the enigma and placed their hopes in a large 4-foot-diameter spirit magnifier then under construction for Jean Charles de Trudaine. The question thus engaged the scholarly community in a race for instrumental performance that certainly attracted the curious, but excluded amateurs from experimental research. Contrary to what most chemistry textbooks claim, neither Lavoisier nor Macquer succeeded in grasping the nature of diamond. They only demonstrated its combustibility, which they compared to that of coal. These disappointing results led the Academy to lose interest in the diamond question.

But Louis-Bernard Guyton de Morveau, chemist and lawyer in the Dijon parliament, took up the question of diamonds in a long and laborious series of experiments conducted from 1781 to 1807 thanks to a few diamonds recovered from an English ship during the Anglo-French War (1778–83). Some of these experiments were conducted at the École Polytechnique, created during the French Revolution, where Guyton was professor of mineralogy. It was Guyton, not Lavoisier, who clearly established that 'diamond is pure carbon, the pure acidifying base of carbonic acid'.[6] As early as 1786, in the article 'Air' in the *Dictionnaire de chymie de l'Encyclopédie méthodique*, Guyton states that if the combustion of diamond produces carbonic acid gas, and if this product has the same weight as the air and diamond that have disappeared, then 'there would be no way of refusing to accept the consequence that diamond and carbon are two identical beings'.[7] This, in 1786, is one of the very first appearances of the word 'carbon'.

A creature of nomenclature

In fact, Guyton was an inventor of words. For years, he had been working on a reform of chemical nomenclature,[8] which he finalized with three Parisian chemists: Lavoisier, Claude-Louis Berthollet and Antoine de Fourcroy. In the spring of 1787, they presented their *Méthode de nomenclature* to the Academy.[9] The term 'carbon' (*carbone*) appears in the lexicon of new names and its equivalent in the column of old names is 'pure coal'. The invention of carbon did not come directly from studies of diamond or from joint studies of fixed air. The 'Carbon' entry in the *Dictionnaire de chymie de l'Encyclopédie méthodique* written by Fourcroy for the third volume does not even mention diamond but explicitly refers to nomenclature as the origin of the term:

> The word carbon, used in the chemical nomenclature proposed several years ago by Morveau, Lavoisier, Berthollet and myself, and adopted today by the majority of chemists, was taken from the Latin word *carbo, carbonis*; it is used to designate the pure matter of coal. . . . Now, as chemists, when using common coal in their experiments, almost always take it up only because of the pure carbonaceous matter that it contains, and to carry this matter only by virtue of the attractions that it exerts on other bodies to which they want to unite it, they had to look for a name that would express especially this carbonaceous base, this true principle of coal that produces all the particular effects that distinguish it.[10]

Does this creature of nomenclature really identify with what we now refer to as carbon? The carbon of 1786 is presented as a carbonic acid radical; its distinctive property being combustion, it suggests a strong affinity for oxygen, 'which dissolves it into an elastic fluid, forming carbonic acid'. So strong is this affinity that carbon is capable of expelling the acidifying principle (oxygen) from almost all bodies that contain it. But since no other body can in turn separate it, it is impossible to isolate carbon:

> If a body were discovered which had more affinity for oxygen, it would be possible to obtain this principle, carbon, separately; then it would be known in a state of purity, and all its properties could be studied. . . . It is the case with carbon, as with many simple materials, that we can never hope to obtain it in a state of perfect purity. The moment it separates from a compound of which it is a part, it suddenly enters another; it absorbs all neighbouring bodies; it unites closely with the fixed salts that it finds within its reach; it turns back into coal . . . but in spite of this difficulty one can at least in thought isolate the carbon that makes up the greater part of coal.[11]

Carbon is thus designated first as a simple body, then as a principle, then a radical, but never as an element. It retains many of the features of traditional principles, those vehicles of properties that they confer on the compounds they form. Fourcroy explains the meaning of the word by saying that 'alchemists . . . would have used the expression "the soul of coal, the essence, or the spirit of coal"'. But a reformed nomenclature requires unambiguous language, with simple names for simple substances evoking the properties of the designated body.

> Thus, the word carbon, which has all these characters, clearly expresses a body contained in coal, making it its base, the principal character of coal, even constituting it as such, and its difference from the word coal, of which some people have reproached us because in general they have not understood us, is one of its principal advantages, since it is intended to make it clear that what is in question, with this word, is a body which has very significant relations with coal properly so called, and yet is not coal . . . a name which belongs especially to coal as clean, pure matter, and which, by obliging to consider it in whole coal, could lead thinking to isolate this matter from the whole mass of coal and to follow it in all the combinations which are proper to it.[12]

Carbon thus remains as close as possible to coal. Moreover, in enumerating the properties of carbon, Fourcroy considers that 'there is almost no doubt that the essence of carbon is to be black; it even seems that this body is the only one in nature that truly has this colour, and that can communicate it to others'.[13]

Coal's footprint

So it is coal, more than mephitic air or diamond, that provided the template for thinking about carbon. It is the stamp on which the word 'carbon' was embossed. And this is not surprising if one considers that chemists, since the medieval period, when they were designated as 'philosophers of fire', had acquired an intimate knowledge of coal in their daily practice. As the 'Coal' entry in Diderot and d'Alembert's *Encyclopédie* pointedly stated:

> [I]t provides the Chemist with the most ordinary and convenient fuel for the fire he uses in most of his operations. This coal chosen must be hard, compact, sound and dry; it must also be whole perfect coal, or, which is the same thing, not mixed with fumaroles: this choice is important mostly for the convenience of the artist.[14]

Moreover, eighteenth-century chemists, most of whom were involved in arts and crafts, were familiar with coal because it was used in the manufacture of a variety of products. Charcoal was first used by black-smiths and glassmakers, who worked at high temperatures that could not be reached by burning wood alone, even with the help of large bellows. For coal and charcoal significantly differ. Starting with 'the natural and the artificial; these two substances have almost nothing in common but the colour that comes out of them',[15] says the 'Coal' entry of the *Encyclopédie*. Burning wood to manufacture charcoal was a complicated art that fell under the mechanical arts, while mineral coal fell under natural history. Chemists classified this artificial coal as earth, like Wilhelm Homberg, or as sulphur, like Herman Boerhaave, and all admitted that its composition differed according to the place. When Lavoisier and Guyton, as well as Smithson Tennant in England,[16] tried to show that by burning an equivalent mass of diamond and coal one obtained an identical volume of fixed air, their numerical values all differed from one another, mostly due to the different kinds of coals they had used. More than diamond, which was too pure, it was ordinary coal that eluded mastery, even though people thought it held the key to the nature of diamond.

Certainly, most chemists assumed that coal was a body composed of phlogiston and an oily substance, or something else. But once Lavoisier had banished phlogiston, the authors of the *Méthode de nomenclature* seemed to falter a bit. Guyton first introduced carbon into the category of simple bodies under the name of 'coal' ('*le charbon*'), because in his view the radical that forms carbonic acid was 'pure carbonaceous matter'.[17] For him, 'carbon' was just a more precise, scholarly term for coal:

> To make the name of this radical even more precise, by distinguishing it from coal in the vulgar sense, by isolating it by thought from the small portion of foreign matter that it usually contains, & which constitutes the ash, we adapt to it the modified expression of carbon, which will indicate the pure, essential principle of coal, & which will have the advantage of specifying it by a single word, in such a way as to prevent any equivocation.[18]

Thus coal, diamond and carbon were presented as degrees on a scale of purity. Coal is *almost* pure carbon, while 'diamond is the purest carbon, the pure acidifiable base of carbonic acid'.[19]

Word battles

But more precisely what was the nature of this carbon, the ultimate degree of purity of several more or less familiar material bodies? The chemists who wrote the *Encyclopédie méthodique*, and were responsible for spreading the news, did not announce its invention as the birth of a new body. It was above all a word, a simple word that made it possible to connect familiar but very diverse things. And since coal was more a part of their daily experience than diamond, they tended to colour carbon black.

Naming substances is the perpetual bane of chemists, who are anxious to find an order among substances for teaching purposes. Hence there is a considerable effort of rationalization, in the precise sense of a search for invariants in the indefinite variation of the substances with which they are dealing.

Carbon in the late eighteenth century was still not linked to graphite. Pencil leads were indeed made from this substance, but it was then identified as plumbago, or 'lead mine'. The analogy between coal, graphite and diamond required, in addition to chemical research, the study of the geometric form of crystals, i.e. crystallography, which was in full swing by the end of the eighteenth century.[20] The geometric description of crystals and the definition of each substance by the angle formed by the crystal led to a classification of crystals according to their shape and based on the distinction resulting from the work of René-Just Haüy between 'elementary molecule' and 'integral molecule'. The invisible elementary molecules enter in the formation of the integral molecules characterized by a shape – parallelepiped, triangle or tetrahedron – that is stacked regularly to constitute the crystal. This distinction is not unlike the one introduced by Amedeo Avogadro in 1811 and that chemists later renamed atom and molecule. However, for decades the vocabulary that would allow coal, graphite and diamond to be defined as three different states of the same carbon was not stabilized. Plumbago and anthraxolite were considered as more or less oxidized intermediates between diamond and coal, and coal as an oxide of the second-degree precursor of carbonic acid, which would be completely oxidized carbon. If 'diamond is pure carbon', coal remains impure: it is a diamond protoxide (i.e. containing oxygen), as Guyton asserted in 1812; or, conversely, 'diamond is pure coal', as Joseph Gay-Lussac maintained in 1818. Nineteenth-century chemists, confronted with the multiplicity of forms and properties that solid bodies can present, never stopped arguing about the way to designate the relations between the same and the other.

The word 'allotropic', which means 'in another trope', i.e. with another manner or quality, or even 'another *behaviour*' (and not of another 'form', as is often said), was introduced by the Swedish chemist Jacob Berzelius, who played the role of band leader of the international community thanks to his Annual Reviews.[21] He had already coined the word 'isomer' (same parts) to designate bodies of identical composition but different properties. He presented allotropy as a species of the genus isomerism, concerning elementary bodies:

> Some, he remarks, have the property of affecting, under the influence of certain as yet undetermined circumstances, an external state, or different forms, which they appear to preserve in several combinations. . . . We call this fact allotropy.[22]

Even if three examples – sulphur, carbon and silicon – shed some light on this definition, it remained too imprecise to distinguish polymorphic bodies (which occur in different crystalline forms) from isomeric bodies (which occur in different states, e.g. crystalline or amorphous).[23] It did not satisfy the chemist Alfred Naquet. In his view, allotropes were like different races, isomers like different species.[24] With regard to carbon, on the one hand, he emphasized its polymorphism – the multiple forms of coal, lampblack, animal black, etc.; on the other hand, he related the multiple 'allotropic forms' of carbon to differences in state: amorphous in ordinary coal, octahedral crystalline in the case of diamond and rhombohedral in the case of graphite.[25] By 1860, the puzzle of the allotropic forms of carbon was still unresolved. In a debate with the chemist Benjamin Brodie, Naquet cautiously concluded:

> It is highly probable that graphite and diamond are two clearly separate allotropic states of one and the same body. But does diamond differ from ordinary carbon other than by the very fact of crystallization? This is a difficult problem to solve, since ordinary carbon, like diamond, is insoluble in all solvents. The differences in colouring and the property that diamond

possesses of returning to the amorphous carbon state when the temperature is greatly increased could alone make it possible to lean in favour of this hypothesis.[26]

Thus carbon as a single chemical entity, existing under various phenomenological appearances, slowly, painfully, came into existence. Three devices contributed to the process of emergence of this abstract notion: two instruments and one linguistic tool. We have seen that scales at work in the identification of fixed air provided a powerful means of tracking and tracing what passes and circulates from one body to another by accounting for what enters and leaves the black box of material transformations. We have seen that the burning mirror – a large and expensive device allowing combustion without coal – played a crucial role in establishing the combustibility of diamond and bringing it closer to ordinary coal. Finally, the reform of the chemical language opened up an opportunity to denominate, i.e. to substantiate, what had been accounted for through the balance sheets of chemical reactions, then identified and characterized by way of trials and tests.

But the substance named carbon in the nomenclature of 1787 was still soot black and impregnated with coal, because it was discovered and identified mainly in combustion and calcination operations linked to the arts and crafts as much as to the investigation of nature. A considerable effort of concept creation and distinction was still required to recognize diamond, coal and graphite as three concrete embodiments of the same element: carbon. This was only possible through the invention and refinement of the allotrope concept. Without it, to substantiate carbon, to abstract it from black coal and pure diamond, was simply inconceivable. These linguistic devices – painstaking and much less spectacular than the great experiments with diamond – nevertheless played an essential role in the invention of carbon.

4

An exemplary element

Gustave Flaubert immortalized Bouvard and Pécuchet as two gentlemen who were eager for modern knowledge. They learned about chemistry by buying textbooks:

> To learn about chemistry, they procured Regnault's textbook and the first thing they found out was that 'simple bodies might actually be compounds'. These objects are divided into metalloids and metals – a distinction, said the author, that 'is in no way absolute'. . . . And so they turned to a less demanding work, Girardin's textbook, from which they gained the certainty that ten litres of air weigh one hundred grams, that pencils do not contain lead, and that diamonds are merely coal.[1]

It may not have been Flaubert's intention to show that chemistry books were full of nonsense, but he took a malicious pleasure in casting doubt on a science that was well established in bourgeois society and widely disseminated through courses and treatises. He particularly targeted textbooks, those marvellous instruments of control that proliferated during the nineteenth century when chemistry became a respectable and highly praised scientific discipline.[2] Chemistry participated in the production of wealth by providing more and more so-called 'artificial' compounds, the keys to industrial and agricultural development. Chemists became professionals in their own right, sanctioned by diplomas at the end of a course of study. The discipline was taught in secondary schools as well as in universities, which trained experts for industry, health, safety, urban planning, the courts, etc.[3] Chemistry disciplined citizens by flushing out frauds and other crimes thanks to its analytical techniques, at the same time as it tried to discipline the burgeoning number of products that it generated as it progressed.

Carbon was a central actor in this vast disciplinary enterprise, even if it played a somewhat ambivalent role: sometimes a model student, sometimes a rebellious one. It required special treatment because of the multiplicity of its states and forms, as revealed in the previous chapter. Carbon raised a central question: how does one define the identity of a body? Is it the same or other? Is it unique or multiple? Roald Hoffmann, poet and chemist (winner of the Nobel Prize in 1981), emphasizes the emotional density attached to these very abstract questions:

> It is possible to answer the question, 'What do I have?'... But why is that question interesting? Because the question of identity, and *our* identity, shaped in childhood in a complex dance of bonding and separation, matters deeply to us. The processes of nature connect with the interior world of our emotions.[4]

A textbook example

Professional chemists learn their trade through what Thomas Kuhn calls a 'paradigm', i.e. a 'disciplinary matrix' of concepts, values and norms transmitted through textbooks. Such manuals follow a pedagogical progression that has nothing to do with historical order; they offer typical examples, *exemplars*, as well as exercises.[5] What about the textbook consulted by Bouvard and Pécuchet?

Henri Victor Regnault's *Cours élémentaire*, published in 1840, attempted to discipline the multiplicity of material bodies that a good student must master by following the didactic method vigorously upheld by Lavoisier. Proceed from the simple to the compound, i.e. start by defining the simple bodies, draw up a list of them and then characterize the compounds by the nature and proportion of the simple bodies that form them. In the passage to which Flaubert explicitly referred, Regnault presented the simple bodies as follows:

> We do not wish to assert by this that these bodies are really simple; it is very possible that the future progress of science will enable us, progressively, to carry out decompositions which have resisted our present means; and that then a certain number of the bodies which we regard today as simple, perhaps even all of them, will be considered as compound bodies.[6]

Like most chemists of his time, Regnault used here the classical concept of the element as 'the last term arrived at by analysis',[7] which gives them a provisional character because their simplicity depends on the progress of analytical techniques. With his stinging irony, Flaubert noted a concern raised by such a concept: in a science organized along the axis of simple to compound, where simple bodies are the basis of the whole edifice, this definition installs a radical uncertainty into the building blocks of the construction. Flaubert also pointed out an obvious inconsistency in the classification of simple bodies that governs the organization of most chemistry textbooks at the beginning of the nineteenth century: metals and metalloids or non-metals.[8] Regnault raised doubts about this distinction and emphasized its artificial character by showing that some bodies had characteristics of both classes.[9] He nevertheless adopted this classification after having criticized it, contenting himself with a few detailed refinements.[10] In short, chemistry was a bundle of contradictions that lent itself marvellously to the making of a 'critical encyclopaedia as farce'.[11]

A few years later, Dmitri Mendeleev, professor of chemistry at the University of St Petersburg, wrote a textbook on general chemistry, published in Russian in 1871. In a long introduction, he admitted that the notion of a simple body – i.e. one that had not been decomposed by contemporaneous means – was the basis of chemistry. He began by listing the seventy simple bodies known at the time, in order of natural abundance: first hydrogen (present in water and organisms), then carbon (present in organisms, coal and limestone), nitrogen, etc.[12] However, Mendeleev introduced 'a clear distinction between the conception of an element as a *separate homogeneous substance*, and as a *material but invisible part of a compound*. Mercury oxide does not contain two simple bodies, a gas and a metal, but two elements, mercury and oxygen.'[13] To justify this unintuitive distinction between the simple body and the element proper, he mentioned the case of carbon:

> Besides, many elements exist under various visible forms whilst the intrinsic element contained in these various forms is something which is not subject to change. Thus carbon appears as charcoal, graphite, and diamond, but yet the element carbon alone, contained in each, is one and the same. Carbonic anhydride contains carbon, and not charcoal, or graphite, or the diamond.[14]

Carbon thus offered an enlightening solution to the question of material identity: the element carbon was the variable invariant, the same in the other, the one thing underlying the multiple forms and states of simple or compound bodies. The distinction between element and simple body, which was considered superfluous in Lavoisier's chemistry, centred on analysis, on the back-and-forth between simple and compound, was nevertheless essential for those who sought to bring order to the suffo- cating multiplicity of chemical substances. It allowed chemists to escape the dilemma of the two blocks – metals, non-metals (or metalloids) – into which it was necessary to fit all the elements/simple bodies. The case of carbon was again summoned to undermine these two categories: 'Thus graphite, from which pencils are manufactured, is an element with the lustre and other properties of a metal, but charcoal and the diamond, which are composed of the same substance as graphite, do not show any metallic properties.'[15]

A material abstraction

This conceptual distinction, illuminated by carbon, proved so crucial to the construction of the periodic system of elements – rather than simple bodies – that it appears at the top of the article in which Mendeleev announced his discovery:

> A simple body is something material, a metal or a metalloid, endowed with physical and chemical properties. The idea that corresponds with the expression simple body is that of the molecule. ... By contrast, we need to reserve the name element to characterize the material particles that constitute the simple bodies and compounds and that determine the manner in which they behave in terms of their physical or chemical properties. The word element should summon up the idea of the atom.[16]

Mendeleev took care to define the element very precisely by its relationship with other concepts that refer to concrete physical or chemical entities: the simple body corresponds to the molecule, the element to the atom. The concept of element 'summons up' the idea of atom, but it should not be confused with it. Mendeleev could assume the existence of an element such as carbon even though he doubted the real existence of atoms. All that mattered was the distinction between atoms and molecules, implied by Avogadro's law.[17] This served as a tool for defining the element thanks to the analogy established between two pairs of terms: atom/molecule, on the scale of microscopic entities; element/single body, on the scale of bodies that we can handle. Carbon as element was thus defined by its place in a network of concepts.

So what is carbon? Graphite and diamond are simple bodies; calcium carbonate and carbonic acid are compound bodies (made up of carbon and something else). But all of them are *bodies*, concrete things, which exist in a certain state (solid, liquid or gas), and can be isolated after a process of analysis and purification. In contrast, as an element, carbon cannot be seen, touched or isolated. It is a purely abstract material entity, but can nevertheless be known by an individual characteristic: its atomic weight. For Mendeleev, the atomic weight, which 'belongs neither to coal nor to diamond but to carbon', is the *signature* of the element.[18] And for carbon, this signature is 12.

We can now understand why it was necessary to install such a non-tangible thing, a real metaphysical entity, into a supposedly positive experimental science. It was by freeing carbon from the panoply of concrete substances, from their familiar appearances, from their industrial uses, by disregarding its physical and chemical properties and even the idea of the atom, that Mendeleev was able to focus on a single signature of carbon, its atomic weight. The detour into the abstract proved a necessary step for discovering analogies among bodies and finding the general law that governs their relationships: the periodicity of properties as a function of mass.

In his famous table, Mendeleev ordered the elements by periods and columns (or groups). The elements were aligned in order of increasing atomic mass and grouped in columns when they show similar chemical behaviour. Carbon was listed as the head of a column. It was the main representative of a 'family' in which silicon, then titanium and, a little later, germanium were listed below it in the table. The periodic table offered a systematic way of articulating the diverse to the analogous without reducing it to the same, thanks to the periodic law that orders the elements according to their atomic weight: as the latter increases, the properties of the elements tend to repeat themselves periodically. Although individual, the elements form a system.

Carbon thus provides the model – the *exemplar* – of the concept of the element involved in Mendeleev's table. The periodic system is neither a collection of concrete substances (black, red or green, as some graphic representations suggest, which include in each box a photograph of the corresponding simple body), nor a collection of atoms (as current graphic representations of the periodic system suggest). This system is based on an abstract notion that is not intuitive. It is thanks to this move towards abstraction that Mendeleev was able, on the one hand, to identify a general law that embraces the multiplicity of elements that had previously been uncontrollable due to the confusion between elements and simple bodies; on the other hand, to predict unknown elements, since the existence of a simple body is not, in the strict sense, foreseeable, insofar as a simple body only exists at the end of an analytical chemical operation that isolates it as such.

If the concept of the element was operative in the construction of the periodic system, in turn, the periodic system makes it possible to specify the kind of entity that the chemical elements are. Mendeleev, who firmly rejected any hypothesis of the unity of matter, insists on the individuality and plurality of elements. Carbon is a unique and singular entity, individualized by its atomic weight C = 12. Its individuality is shaped in the network of the table constructed on the basis of the periodic law. It is enriched by its relationships with all the other elements distributed in the columns and periods of the periodic table.

Carbon thus played a decisive role in the construction of the periodic table by offering, thanks to its allotropic forms, an exemplary and telling illustration of the distinction between element and simple body. It helped to define the taxonomy's object and principle.

In so doing, Mendeleev reconfigured chemistry as a whole. The discipline, organized by Lavoisier around the simple/compound axis in a back-and-forth between analysis and synthesis, now focused on the relationships between the observable and measurable properties of chemical bodies and on what determines their interactions. The aim was to relate the phenomenological behaviour of the most diverse bodies to something abstract that is identified by its atomic weight without being confused with the atom. Atoms and molecules are structural units of matter, building blocks. Even if they are invisible to our eyes, they have a concrete existence on the nano scale, just as simple and compound bodies have a physical existence on our scale. But the element that, for Mendeleev, assumes the explanatory burden of individuality is an abstract entity, a material abstraction. Carbon is a thing (*chose*, with an etymological connection to *cause*), an invisible thing known by its effects.

A metaphysical substance

Few chemists nowadays pay attention to the distinction introduced by Mendeleev. For many contemporaries, carbon is the carbon atom that goes into carbon molecules. Yet the distinction between element and simple body saved the periodic system in 1914, when radioactive elements were isolated. Where should these strange things be placed, which are at the same time similar to the known elements and yet different in their behaviour and atomic weight? The fact that the same element could vary in atomic weight cast doubt on the identity of the element. Conversely, the fact that the variation in atomic weight was accompanied by a change in the behaviour of the substance seemed to confirm the notion of an element. What was to be done?

These newcomers posed a dilemma for chemists: either revise the periodic table to make room for them, as suggested by Kasimir Fajans, or revise the definition of elements characterized by their atomic weight. Friedrich Paneth and his colleague George Hevesy considered that radioactive elements were *physically different* but *chemically equivalent*. Combining the distinction introduced by Mendeleev between element and simple body with a clear division between physics and chemistry, they therefore placed the radioactive elements in the same box as the familiar elements, and for this reason called them 'isotopes' (meaning 'in the same place'). Thus Paneth and Hevesy preserved the structure of the periodic table but redefined the element as a substance whose atomic nuclei have the same charge.[19] Hence the official definition adopted in 1923 by the International Union of Pure and Applied Chemistry (IUPAC): 'a set of atoms with the same atomic number (number of protons in the nucleus)'.

From then on, the material abstraction constructed by Mendeleev on the basis of carbon was most often reinterpreted as a metaphysical entity. Georges Urbain, a French chemist who participated in the IUPAC commission in charge of definitions, declared that the element had an 'ideological' character, in the sense that it was only an idea:

There is something distinctly mysterious about this something that is common to a simple body and all its combinations. However, its very existence seems indisputable to us, even though it is only positively there as a mind's-eye view. It is this something that I will provisionally call an element. Because of this the element cannot be monitored by the senses straight away. This notion, the experimental origin of which is indisputable, has nevertheless – and I insist on this point – an ideological character.[20]

Urbain turned the element into a substance, in the etymological sense of what-is-kept-underneath, and answered the question: 'What is conserved in chemical transformations?' His judgement was ambivalent. He suggested that as a substance, the element was a philosophically suspect concept, a wholly metaphysical entity at the heart of an experimental and positive science. However, he recognized that, for want of a better concept, it had an indisputable existence and value from the point of view of chemical practice. On the one hand, it was a creature of reason; on the other, it had an 'existence of its own' whose experimental origin was undeniable. Mendeleev, by contrast, used the term 'substance' in a very general sense to denote 'that which occupies space and has weight, that is, which presents a mass attracted by the earth and by other masses of material'.[21] Thus defined exclusively by primary qualities (extension, gravity), substance had nothing to do with the individuality he attributed to chemical elements. Mendeleev's element was not so much a substance endowed with primary qualities and hidden behind appearances as *a sign of a singular mode of behaviour* of matter.

Paneth, on the other hand, fully assumed the metaphysical character and value of the element for chemistry, or, if you like, its 'metachemical' character.[22] He translated the distinction between element and simple body into German terms that gave the element a new ontological status: the element was said to be *Grundstoff* (basic substance) and the simple body *einfacher Stoff* (simple substance).[23] The basic substance was indestructible, present in compounds and simple bodies, and devoid of secondary qualities. The simple substance was only an occurrence of the basic substance, not combined, and accessible to the senses. It related to naïve realism. These terms recall the Aristotelian notion of *hupokeimenon* (ὑποκείμενον) which constitutes the fundamental level of reality underlying the particular beings encountered in experience, the *ek-keimenon* (εκ-κείμενον). Carbon as *Grundstoff* thus answers Aristotle's question 'what is it?' (*to esti*). And for Paneth, as for Aristotle, it is both a concrete singular being, a 'this' that can be seen and touched (*ek-keimenon*), and the substratum that underlies it (*hupokeimenon*). But Paneth gave a Kantian twist to these concepts by stressing the transcendental existence of basic substances to clearly distinguish the element from the atom. He went so far as to criticize Mendeleev for having drawn a parallel between the distinction between element/single body and atom/molecule. In his view, the 'basic substance' was non-individualized, devoid of any quality, and independent of atomism, even though the atom seemed to him to be useful in providing a visual translation of how the elements persisted in the compound. Paneth nevertheless concluded that by choosing the element as the central unit, Mendeleev clearly asserted the originality of chemistry, as opposed to physics.[24]

In summary, this chapter has suggested that the chemical existence of carbon is a life of asceticism and purification. In a highly disciplinary environment, carbon was cleansed of the coal slag, diamond slivers and mephitic vapours that tarnished its nascent state. In the course of modest conceptual work by Mendeleev and others after him, carbon was detached from mundane things and became a pure abstraction, a creation of the mind. It then obediently fitted into the boxes of a system, was given the number 12 (its atomic weight), then 6 (its atomic number). Like the other elements, carbon was caught in the mesh of a system governed by a law, the periodic law, which Mendeleev did not hesitate to place alongside Newton's law of universal gravitation and Kant's moral law.[25]

This formidable regimentation device was deployed in the context of a discipline, in the sense of a subject taught at school, which requires that order be introduced into the teeming multiplicity of material beings invented by chemists with their techniques of analysis and synthesis. Because they had emerged outside the scholarly world designed to shape citizens in touch with their natural and social environment, Bouvard and Pécuchet were unable to apprehend the subtleties of the chemical existence of material bodies that constituted their scholarly mode of existence. After trying to experiment a few times in their kitchen, they quickly abandoned the field of chemistry.

5

Carbon liberates itself

So far we have first told the story of the invention of carbon in the context of an analytical chemistry that aimed to break down compounds to the ultimate degree, until a list of several dozen simple bodies was established. Then we recounted how carbon became the model element that served as a guide for Mendeleev to reorient chemistry by revealing a general law. Analytical chemistry, which tried to explain the compound by the simple, gave way to an elementary chemistry that found the key to compounds and simples in the abstract reality of chemical elements.

One among others

These two research orientations are based on a common postulate, or rather a raw and stubborn fact: the existence of a multiplicity of individual elements. This multiplicity is indefinite insofar as it depends on the progress of analytical techniques; we have gone from thirty-three simple bodies in Lavoisier's *Traité* to seventy in Mendeleev's. To manage and discipline this multiplicity, nineteenth-century chemists developed two alternative strategies: either they tried to reduce the multiplicity of simple bodies until they reached a primitive matter, a primordial element – according to the hypothesis formulated by William Prout at the very beginning of the century – or they tried to subsume the irreducible diversity of elements under a general law, following Mendeleev's preference. These two paths form the philosophical landscape of what Bachelard called the *coherent pluralism of modern chemistry*.[1]

In such a context, carbon appears as an individual among dozens of well-identified elements, characterized by their atomic weight, and arranged in the cells of the periodic table. Nothing *a priori* predetermines this element, at the top of a column in the Mendeleev table, to a unique destiny. It is certainly exemplary, but why should it be exceptional? Yet in the middle of the nineteenth century, this individual, lost in the crowd, became singular. Like an epic hero, it emerged from the mass, tearing itself away from the multitude by capturing the attention of chemists to the point of demanding its very own chemistry. Carbon was carving out a kingdom for itself in the land of chemistry, to the point of recomposing its entire landscape.

The biography of carbon thus leads us to rethink how a field of research is constituted, how a discipline is structured and subdivided into sub-disciplines. The birth of organic chemistry – in the modern sense of carbon chemistry – has been told many times. It often takes the form of a battle of giants, a series of theoretical controversies between great heroes: Jacob Berzelius, Justus von Liebig and Jean-Baptiste Dumas in the 1830s and, in the next generation, Charles-Adolphe Wurtz, August von Kekulé and Marcellin Berthelot.[2] It is sometimes described as a silent revolution led by more discreet figures, such as the German chemist Hermann Kolbe.[3] Most of these accounts are garnished with power plays by influential figures in advantageous institutional positions who dispatch bold innovators like Auguste Laurent and Charles Gerhardt to the provinces.[4] These leadership struggles are spiced with nationalistic ingredients exacerbated by the political tension between Germany and France after 1870.[5] More rarely, it is instruments and practices that come to the fore.[6] Yet in the majority of these loud and boisterous narratives, there is complete silence on the role of the star of this story, carbon.

Let us try to revisit this episode by focusing on the element responsible for this restructuring of chemistry. How does carbon itself help to promote it? What does it suggest? What perspectives does it open up? What opportunities for action does it offer?

Two or three chemistries?[7]

As the historian Ursula Klein has rightly pointed out, the advent of carbon chemistry created a new ontology established through formulae and classifications.[8] In the eighteenth century, three chemistries were distinguished. The treatises were divided according to the major kingdoms of nature: mineral, vegetable and animal. Even if the theoretical basis were common to all three specialities, they differed in their practical aspects: mineral chemistry was geared towards mining, glass and metalworking; plant chemistry was geared towards pharmacy; animal chemistry was an auxiliary science for physiology and medicine. Chemists and pharmacists developed sophisticated analytical techniques to isolate the 'immediate principles' of plant and animal substances, which are often the active ingredients sought for dyeing or pharmacy (for which we still speak of 'active principles'). Immediate analysis, which proceeds in stages, sheds light on the *constitution* of the compounds (i.e. the principles responsible for their agency), while so-called 'elementary analysis', which aims directly at revealing the ultimate elements, delivers the *composition* of plant substances (i.e. the parts of which they are composed). This latter analytic technique, promoted by Lavoisier in his *Traité élémentaire de chimie*, led to the demonstration that the substances produced by plants and animals are mainly composed of carbon, hydrogen, oxygen and nitrogen (especially in animals).

In 1800, when Antoine Fourcroy published his voluminous *Système de chimie*, carbon was listed as one of the four elements of the living world.[9] This was already pride of place, but it remained essentially honorary. None of these elements, nor the group of four taken together, leads to a profound transformation of chemistry. To name and classify the immediate principles of plants and animals, Fourcroy uses heterogeneous criteria: physical and chemical properties, therapeutic properties or their concrete origin.

Until the 1830s, oxygen enjoyed a privileged status in the multitude. Lavoisier attributed multiple roles to it: it overtook phlogiston in combustion and calcination, and was the principle that generated acidity. Although this last function – let's not forget this was the origin of its name – was quickly contested by several chemists and definitively abandoned in the 1820s, oxygen offers a taxonomic principle for classifying material diversity: the degree of affinity of simple bodies for oxygen is used as a reference for classifying them in Louis-Jacques Thenard's *Traité de chimie*. With its four volumes totalling more than 2,700 pages (and more than three thousand in the sixth edition), Thenard's treatise provides a systematic exposition of all chemical knowledge. It served as a reference work in France from 1813 to 1836, and as the official model for all chemistry textbooks. Here is why he focuses on oxygen:

> Before revealing the formation of the names of compound bodies, it is
> necessary to say that these bodies are not as numerous as one might imagine;

that far from there being an infinity of them, there are not as many as there are possible combinations 2 to 2, 3 to 3, etc. between simple bodies. In fact, the compound bodies known up to now result for the most part: (1) from the combination of oxygen with each of the combustible bodies; (2) from the combination of a simple body united with oxygen with another simple body united with oxygen; (3) from the combination of two, three simple bodies together, rarely four; (4) from the combination of oxygen with hydrogen and carbon, the principles which constitute vegetable matter; (5) from the combination of oxygen with hydrogen, carbon and nitrogen, and sometimes phosphorus and sulphur, the principles which constitute animal matter.[10]

Thanks to this 'exaggeration of the role of oxygen',[11] Thenard proposed a classification based on a single criterion – reactivity with oxygen – which works wonders, in particular, in the metals group.[12] Admittedly, oxygen's priority is diluted a little in water, a case of putting hydrogen first, but there is still no mention of carbon. Its presence is attested in all the immediate principles but it keeps a low profile.

In the last edition of his *Traité* in 1835, Thenard renamed the volume devoted to plant and animal chemistry *Chimie organique* [Organic Chemistry]. This semantic shift accentuates the tendency of chemistry, already evident in the reform of nomenclature, to detach itself from the natural origin of substances and consider only their composition or behaviour in the laboratory. This is an important ontological change and it tormented chemists worried about the identity and unity of their discipline. This change posed serious problems indeed. For if the reference to the vegetable and animal kingdoms is removed, mineral chemistry ends up redefined in negative terms as 'inorganic chemistry'. Inorganic chemistry, which had a logical priority due to the reduction of all organic compounds to four elements – carbon, hydrogen, oxygen and nitrogen – is now subordinated to the definition of organic chemistry. Moreover, the division between organic and inorganic chemistry contradicts the traditionally analytical logic of chemistry, which is attached to explaining the compound by the way of the simple. There is therefore no *a priori* justification for organic chemistry to break free. It is the same elements making up water, ammonia, carbon dioxide, starch, urea and fibrin.

The term 'organic chemistry', introduced by Jacob Berzelius in 1808, was therefore slow to catch on.[13] The division between mineral and organic chemistry was violently criticized in 1835 by Auguste Comte in the chemistry lessons of his *Cours de philosophie positive*.[14] Comte proclaimed the unity of chemistry and relegated the study of plant and animal compounds to physiology. In this, he was faithful to Berzelius, who extended to organic compounds his dualistic theory of the formation of inorganic compounds by neutralization of electro-positive and electronegative elements or radicals. Berzelius introduced symbols (initial of the Latin name, e.g. C for carbon) for each element and proposed two types of formulae: 'empirical formulae' – e.g. C^2H^6O for alcohol (the numbers were superscripts at the time) – simply indicating the nature and proportion of each element in the compound; and 'rational formulae' providing a hypothetical interpretation of the constitution of the compounds. As Klein points out, these formulae are 'paper tools' that have a heuristic power comparable to laboratory instruments. They can be manipulated with paper and pencil and various interpretations of the constitution of organic compounds can then be tried out in experiments.[15] Berzelius proposed two rational formulae for alcohol: $(2C + 6H) + O$ and $(2C + 4H) + (H^2O)$. In all cases, the dualistic interpretation of organic compounds is analogous to that of salts (constituted by the union of an acid and a base). Thus, for Berzelius, organic chemistry is constructed in the image of inorganic chemistry. How, then, could carbon chemistry become autonomous and free itself from the model of salts?

A quartet of elements

A first answer lies in the surge of compounds coming into existence. Thanks to the combined progress of immediate analysis and elemental analysis, the number of compounds made up of carbon, hydrogen and oxygen increased fivefold between 1789 and 1835 (to which were added alkaloids, nitrogenous compounds). The *kaliapparat* invented by Justus von Liebig in 1830 contributed to this inflation of organic compounds.[16] The apparatus made it possible to determine the carbon content of a huge number of substances in large samples in a single operation.[17] The number of carbon compounds or carbides multiplied to such an extent that, in the 1840s, chemists felt lost in a jungle without a map. Auguste Laurent compared it to a labyrinth:

[W]hen one sees that, with a simple hydrogen carbide and chlorine, one has been able to make a hundred compounds, and that, with these, one can make a great number of others; finally, when one thinks of the absence of any system, of any nomenclature, to classify and denominate this multitude of entities, one wonders, with some concern, if it will be possible, in a few years, to find one's way through the labyrinth of organic chemistry.[18]

By the middle of the nineteenth century, seven to eight thousand carbon compounds had been identified – alcohols, ethers, aldehydes, etc. – to which a formula had to be assigned. Obviously, empirical formulae are insufficient, but what is the meaning of rational formulae? Do they indicate as many degrees of the same composition or the structure of an indefinite number of compounds?

A second answer relates to this question: what kind of music does the quartet of elements – carbon, hydrogen, oxygen and nitrogen – play? How do they interact? In Marcellin Berthelot's view, the synthesis of organic compounds from the four basic elements must unlock the secret of their internal arrangement. Just as, by way of successive decompositions, one passes from starch to sugar, from sugar to alcohol, from alcohol to hydrocarbons and then to the elements, synthesis, marching in the opposite direction from analysis, must proceed by degrees, from simple to complex. In Berthelot's world, the empire of carbon compounds is built up in successive stages and the formulae must reflect each stage of the synthesis. Thus, after synthesizing acetylene (C_2H_2) with an electric arc, Berthelot derived other hydrocarbons, notably benzene (C_6H_6), which he wrote (C_2H_2) (C_2H_2) (C_2H_2) and which he assumed was formed by combining three acetylene molecules. His 'generational formulae' focus on the formation of carbon compounds. They give priority to the genesis and functions of the molecules. And Berthelot ordered the jungle of carbon compounds by distinguishing eight functions through a hierarchy of degrees of composition:[19] first the hydrocarbons formed from the two elements carbon and hydrogen (acetylene, formene, ethylene-benzene, etc.); then the alcohols by addition of oxygen; then the aldehydes, formed from the alcohols by loss of hydrogen; the acids (formic, acetic, tartaric, etc.), formed again from the alcohols by loss of hydrogen; and finally the acids (acetic, tartaric, etc.), formed from the alcohols; then the compound ethers, formed by the union of alcohols and acids; the alkalis, formed by the union of ammonia with alcohols or aldehydes; the amides, formed by the union of ammonia and acids; and finally the metallic radicals, formed by the reaction of metals with certain ethers. By methodically going through all the degrees of composition, Berthelot was promising to reach the point of creating the products of organized beings, setting up the synthesis of urea as a pioneering event of the feats to come.

> Living things themselves are now conceived, from the chemical point of view, only as a sort of laboratory where material principles are assimilated, eliminated and transformed ceaselessly, according to invariable laws that analysis strives to penetrate. It is even more surprising if we think that the immediate principles of living things, the first terms isolated by chemical analysis, can in turn be destroyed by a subsequent analysis and reduced to three or four elementary bodies, similar to those revealed by mineral analysis. How little these elements resemble the materials which supply them by decomposition! Of the four simple bodies that constitute living things, three are gaseous, namely oxygen and nitrogen, the elements of air, and hydrogen, a constituent part of water; while the fourth is solid and fixed: it is carbon, the most characteristic of all the elements that contribute to the formation of organic substances. These four fundamental simple bodies, combined with small proportions of sulphur, phosphorus and various other substances, are the only elements that nature uses in the formation of the infinite variety of vegetable and animal substances. Their combination gives rise to millions of distinct and defined substances.[20]

Within the quartet of elements characteristic of life, carbon, which is 'solid and fixed', stands out for the stability it offers to the formation of successive layers of organic substances. However, nothing authorizes it to claim a chemistry all of its own. So while Berthelot was the founder of organic chemistry, he was opposed to the division of chemistry into two separate fields, organic and inorganic. He insisted on this point: organic chemistry does not designate a *field* (the chemistry of living organisms) but a *method*, synthesis. As a chemical method, it is concerned with the formation of the immediate principles of living things but is not concerned with the formation of the cells, tissues and organs according to which these principles are assembled in the living things themselves[21]

– a matter it leaves to physiology, just as inorganic chemistry leaves the formation of rocks and soils to the natural sciences. Secondly, the elementary bodies of organic chemistry are no different from those of inorganic chemistry, so that the latter has nothing 'inorganic' about it, just as there is no organic matter strictly speaking. It is therefore necessary to resist the division of chemistry into mineral and organic kingdoms, and to stick to the affirmation of a complementarity and symmetry between analysis and synthesis. Thus the quartet of elements in organic chemistry plays the same music as the chemistry of salts. Nor is carbon playing its own special tune.

An exchange centre

However, carbon chemistry liberated itself from the logic of analysis by another route. It began with an observation made by Joseph Gay-Lussac in the context of a study of the action of chlorine on oils and waxes,[22] confirmed by Jean-Baptiste Dumas during an evening at the Tuileries. The candles gave off a pungent smoke in which Dumas spotted hydrochloric acid and assumed that it came from the bleaching of wax with chlorine, and thus from a *substitution* of the hydrogen in the wax by chlorine.

Dumas compared this phenomenon with other cases observed by Michael Faraday in the action of chlorine on ethylene, by Justus von Liebig and Friedrich Wöhler on benzaldehyde, and by himself on turpentine essence. From this he derived a rule of thumb for substitution: when a substance containing hydrogen is subjected to dehydrogenation by the action of chlorine, bromine, iodine, etc., for each hydrogen atom it loses, it gains a chlorine, bromine or iodine atom.

In a voluminous article from 1834, 'Considérations générales sur la composition théorique des matières organiques' [General considerations on the theoretical composition of organic matter], Dumas takes seriously the observation that among the numerous combinations of carbon, many cannot be interpreted within the framework of Berzelius's dualist theory. Carbon makes a mockery of the electrochemical theory. If one assumes two binary compounds in opposite electrical states in organic compounds (on the model of salts), then it must be assumed that the carbon is electropositive in the acid and electronegative in the base.[23]

This dualistic view was further shaken by the evidence of substitutions of chlorine atoms (electronegative) for hydrogen atoms (electropositive) in carbon compounds. In Dumas's view, this was enough to render the division between organic and inorganic chemistry ineffective, in favour of a completely different vision based on different ontological choices, while at the same time asserting an identity unique to chemistry.

> In my opinion, there is no such thing as organic matter, that is to say I only see, in organized beings, some slow apparatus, acting on nascent substances, and thus producing very diverse inorganic combinations with a small number of elements.
>
> Organized beings achieve, regarding the combinations of carbon with the elements of air and those of water, what the great revolutions of the globe have produced for the combinations of silicic acid with the bases that were given to it. On both sides, the same complication. Chemists who maintain that organic substances have something specific in their molecular arrangement, seem to us just as unfounded in their opinion as the mineralo-gists who wanted to see in minerals something other than ordinary chemical species.[24]

Carbon is thus seen as an exchange centre where substitutions are constantly taking place, an immobile orchestra conductor who slows down, distributes or even fixes the course of the elements revolving around them. Dumas suggested a strategy for finding one's way through the labyrinth of organic compounds: it is about multiplying analyses, not merely analysing; about 'carefully following the study of reactions in all their details and linking the two classes of facts by an overall view'.[25] Recentring or even fixing chemistry on carbon as an active substitutor makes it possible to *follow the reactions*. A little later, in a lecture at the Paris School of Medicine in August 1841, Dumas gave a glimpse of a vast cosmic trade where carbon sees itself exchanged, burned, assimilated, fixed, rejected, circulating ceaselessly between plant, animal and mineral beings. This incessant trade tends towards equilibrium, hence the title of that lecture: 'Statique chimique des êtres organisés' [Chemical statics in organized beings]. Fix carbon, then, and all of chemistry comes into your equations; set it in motion again and it is the general economy of nature that strikes the right balance.

More than the analogy of composition, carbon leads us to favour the analogy *of reactions*, which should serve as a guide to order the teeming multitude of substances. While acknowledging that a general theory of organic combinations 'is not something that can be proposed today',[26] Dumas conjectures that organic chemistry will dictate its rules to inorganic chemistry, thus overcoming the arbitrary divide in chemical science. In any case, it is by diverting the attention of chemists from the analysis of compounds to substitution reactions that carbon emancipates itself from the tutelage of salt-inspired models.

A measurement standard

As a substitution node, carbon is definitely no longer one among many. On the contrary, it has taken on a role equivalent to that of the balance in chemical equations, and in so doing, it draws all chemistry to itself. This is how it was able to be of huge service to the system of atomic masses and to the standardization of fundamental chemical values. Carbon, once the *exemplar* of the element concept involved in the construction of the Mendeleev table, became the *standard* for the atomic mass of all elements.

How can atomic masses be quantified when it is impossible to weigh atoms? Since the early nineteenth century, the determination of atomic weights was based on a system of relative values. For example, suppose we know that a hydrogen atom combines with an oxygen atom in a chemical reaction: if we measure, on the scales, that eight grams of oxygen combines with one gram of hydrogen, it is easy to deduce that the oxygen atom has eight times the mass of the hydrogen atom. It is therefore possible to compare the masses with each other as long as we know the reaction formulae, or conversely to find the formulae if we know the masses. However, in order to make progress with this kind of work, chemists need to agree on a standard, a conventional unit to which other masses can be related. John Dalton's first system of atomic masses was based on the standard of hydrogen = 1, because hydrogen is the lightest element and 1 is the simplest number on which to make comparisons. But the table produced by Dalton proved to be incorrect because it was based on the assumption that atoms combine one by one in chemical reactions – the principle that provided the basis for his atomistic hypothesis. Disagreements between discordant atomic weight tables pushed chemists towards other standards. In the days of Berzelius and Thenard, when oxygen enjoyed a privileged position, oxygen = 16 was chosen as the mass standard, as it was supposed to combine with more elementary bodies than any other.

In the twentieth century, radioactive isotopes made a mess of the still shaky system of atomic weights. As discussed in the previous chapter, chemists chose to safeguard the structure of the periodic table by redefining the concept of an element based on the charge of the atomic nucleus, rather than on mass as Mendeleev had done. However, while the atomic number allowed the table to hold together, the masses still fluctuated according to isotopes, usage and disciplines. Thus physicists assign the value 16 to only one isotope of oxygen, while chemists assign the same value 16 to the mixture of varying composition of isotopes 16, 17 and 18 of the element oxygen.

In 1959 and 1960, the International Union of Pure and Applied Chemistry (IUPAC) and the International Union of Pure and Applied Physics (IUPAP) set out to put an end to this unfortunate discrepancy. The chemists and physicists in charge of naming decided to link the atomic mass system to the isotope 12 of carbon. In 1961, an IUPAC commission proposed a unified atomic mass unit (AMU) called the dalton, which corresponds to one twelfth of the mass of an atom of the isotope 12 of carbon, 'unbound, at rest and in its fundamental state', as a 1980 commission specified. Whether such an isolated and impassive carbon atom actually exists is of little importance. Its role is to provide chemists and physicists with an abstract general equivalent, a standard of comparison, or, more precisely, of commensuration, of atomic masses, capable of putting an end to any discord.

As diplomat, carbon provided a common language and standard for chemists and physicists. It allowed chemists to continue, despite the plurality of isotopes, to speak of atomic weight (renamed 'mass', more physically correct) and to use it to predict the proportions of reactants needed for chemical reactions without systematically arousing the suspicion of physicists. It acted as a mediator.

Ten years later, the isotope 12 of carbon was also put forward as the standard definition of the mole or 'quantity of matter', expressed by Avogadro's number, N. This magic number allows chemists to understand the order of magnitude of elementary entities by expressing them in terms of the macroscopic populations manipulated on our scale – or, more simply, to *count* the bodies involved in a reaction on the basis of the knowledge of the formulae and macroscopic masses of the reactants, and even to keep an account of them. Like atomic weight, the 'gram-molecule' value is a dimensionless quantity about which chemists and physicists have been bickering endlessly. Linked to various convergent but imprecise measurements (approaching 60.10^{22}), its value has long fluctuated according to the type of measurement deemed significant to represent it.[27]

However, it was once again carbon that was elected to set the value of Avogadro's number. In 1971, as proposed by the IUPAC, the International System of Units chose to define the mole as the quantity of matter of a system containing as many elementary entities as there are atoms in 0.012 kg of carbon-12 (i.e. $6.022045.10^{23}$). Thus N atoms of ^{12}C have a mass of 12 g. With its number 12, carbon once again acts as a peerless mediator. It acts as a diplomat between chemists and physicists. Between orders of magnitude, between the nanometric scale of atoms and molecules and the macroscopic scale of our material manipulations, it plays the role of converter. A discreet but unavoidable ally, carbon makes life easier for chemists.

6

A relational being

The advent of carbon chemistry came through the emergence of a new concept, atomicity,[1] or the ability to saturate an atom. In the realm of carbon, atoms are defined by their 'combination value', also known as valency or atomicity. These possibilities of entering into relationships are the signature of the carbon atom's identity. This chapter describes the steps involved in uncovering the singularities in the structure of the carbon atom that allow it to form multiple bonds and myriad combinations.

Atomicity

This concept, distinct from affinity – or combinatory power – refers to the character specific to each element. August von Kekulé thus established that carbon is quadrivalent (or tetra-atomic) because it saturates four hydrogen atoms in marsh gas (now called methane, CH_4). One chlorine atom can be substituted for one hydrogen atom to form methyl chloride (CH_3Cl), or three chlorine atoms to form chloroform ($CHCl_3H$), and finally four chlorine atoms can be substituted for these four hydrogen atoms to form carbon chloride (CCl_4).

The two notions of combinatory capacity and substitutability value refer to each other and are anchored in the vision of molecular structures where each atom can be substituted by an atom of another element without changing the structure (see Figure A). However, as soon as they write such formulae, atomistic chemists such as Charles-Adolphe Wurtz point out that they do not indicate the actual position of atoms in space.

> Before proceeding we must warn our readers against an error. Expressions of the kind of which we have just given an example are not intended to describe the position occupied by each atom in space. They indicate the relations which exist between the atoms. ... The hyphens inserted between the atoms

Figure A. Formulae for methane, methyl chloride, chloroform and carbon chloride.

do no more than mark their degree of saturation. They indicate the number and the interchange of the units of saturation, signifying nothing else. Each atom of the quadrivalent carbon is surrounded by four lines, while atoms of the univalent hydrogen are only linked by one.[2]

Structural formulae are first and foremost 'paper tools',[3] graphic techniques for visualizing a possible arrangement of atoms in the molecule as suggested by reactions. They have a heuristic function precisely because they do not seek to represent the internal organization of atoms in molecules but to suggest possible arrangements. Berthelot, who criticizes this freedom of writing by way of signs conceived as representations, denounces the seduction exerted by 'the algebraic ease of their combinations' which replaces 'the direct conception of things, always partly indeterminate'.[4] This is precisely the creative impulse of these formulae, covering the space of indeterminacy opened up by the results of operations carried out in the laboratory. Charles Gerhardt in particular presents his rational formulae as 'summaries of reactions' that say nothing about the actual constitution of molecules.[5] In fact, most of the proponents of atomic theory – Wurtz, Williamson, Gerhardt, Kekulé – remain agnostic about the real existence of atoms. For them, the only thing that matters is the relationships among atoms and their ability to organize themselves into molecules. Their vision of atoms is clearly distinct from the ultimate particles that structure matter. Rather, they are individual dispositions to associate and form bonds. It should be added that, at least for Wurtz, these dispositions are in themselves relative because, like affinity, they vary according to the partners. 'We must therefore consider that the action of atoms is reciprocal, and that in a compound formed by two heterogeneous atoms, the properties of one are influenced by those of the other, both having to accommodate each other in some manner.'[6]

The C–C bond

Carbon is notable for its exceptional capacity to combine. Its achievement, magnificently uncovered by Kekulé, consists in linking with itself to form molecules of several carbons. Indeed, it was already accepted that simple bodies in the gaseous state were formed by a molecule with two atoms of the same nature.[7] Carbon does not do this; it bonds with itself in order to unite with other elements. Thus one carbon atom saturates four hydrogen atoms, but two carbon atoms bind only six, not eight. If all four valencies of carbon were taken up with four hydrogen atoms, we would have two molecules of methane instead of one of ethane. Similarly, three carbon atoms join with only eight hydrogen atoms to form a molecule of propane. It is therefore the list of known hydro-carbons that has guided hypotheses towards the carbon–carbon bond. This exceptional ability of carbon to bind to itself multiplies the possi-bilities of bonding and justifies an autonomous carbon chemistry, as Wurtz clearly grasped:

> The affinity of carbon for carbon is the cause of the infinite variety, the immense multitude of combinations of carbon; it is the *raison d'être* of organic chemistry. No other element possesses to the same degree this master property, this faculty possessed by its atoms to combine, to rivet one another, so as to form this framework so variable in its form, its dimensions, its solidity, and which serves as a sort of fulcrum for the other materials, I mean for the atoms of the other elements.[8]

The carbon–carbon bond was so striking, and so decisive for the future of chemistry, that it was made out to be a revelation. Kekulé claimed to have seen it in a dream, and told a memorial address in 1890 that he had two visions in his career: in 1854, atoms dancing on a London bus revealed the carbon–carbon bond to him; and another vision in front of a fireplace in Ghent in 1862 – the image of a snake biting its tail, recalling the alchemic symbol of the Ouroboros – gave him a glimpse of the structure of benzene.

In fact, the cyclic structure imagined by Kekulé – anticipated by others and reworked to give the regular hexagon well known today – posed a problem, because it involved three carbon–carbon double bonds, in addition to the three single bonds to saturate all the valencies of the carbon. The location of these double bonds should have resulted in two isomers, whereas experience has shown that three are possible. Kekulé's ingenuity in 1872 was to guess that single and double bonds exchange with each other. The resulting oscillation between the two structures is conventionally represented by a circle (Figure B). The benzene hexagon was only obtained after a great deal of work by a number of chemists. In particular, the Scottish chemist Archibald Couper and the Austrian Joseph Loschmidt had proposed approximate hypotheses for the structure of benzene.

But the more or less legendary story of the waking dream testifies to the importance of visualization and mental images in the work of chemists.[9] It is as if carbon revealed to Kekulé one of its special characteristics, which went on to affect the rest of organic chemistry's development. In this way, a founding myth was created for organic chemistry, which at the same time allowed potential rivals to be ruled out and carbon chemistry to be centred on Kekulé.

Asymmetry

A pupil of Kekulé, Jacobus Henricus van't Hoff, continued this intimate dialogue with carbon, and it gave up another of its secrets, asymmetry. This decisive step in the history of chemistry was not legend-making like Kekulé's dream, but instead offered a lovely vignette, a story of two

Figure B. Structure of the benzene molecule.

young chemists, Joseph-Achille Le Bel and van't Hoff, who published the same hypothesis in the same year, 1874, and, instead of fighting over priority, made friends in a vow of 'respectful affection'.[10] Cases of simultaneous discoveries in science are legion. In this particular case, as Le Bel and van't Hoff had worked together in Wurtz's laboratory, the convergence was not entirely unexpected.[11] Yet the approach that led them to the idea of asymmetric molecules was quite different.

When it came to the carbon asymmetry, for Le Bel it was above all a matter of geometry, a spatial distribution of atoms or radicals linked to an atom, such that the two halves of the molecule differ like right and left hands.[12] So the molecule cannot be superimposed on its mirror image; it is a chiral molecule.

Le Bel's approach was entirely geometric, following on from Louis Pasteur's thesis on crystallography. Crystallographers had already identified that certain crystals with asymmetric facets have the power to deflect polarized light to the right or left. This 'rotatory power' is therefore used as an indicator to access the internal structure of compounds. In the 1850s, during a study of tartaric acid and its isomers, Pasteur succeeded in manually separating two types of paratartrate crystals: right-oriented crystals (dextrorotatory) and left-oriented crystals (levorotatory). Only the mixture, known as racemic, of the two types of dissymmetric crystals produces the optical indifference of the paratartrate. Pasteur thus shed light on the question of isomers: these compounds, identical in nature and proportion of their elements, differ in their molecular arrangement.

In 1874, Le Bel set out to determine *a priori* whether or not a substance in solution had rotatory power according to its formula. He sought to predict the physical properties of possible isomers even before synthesizing them. For this purpose, he relied on geometric considerations and, like crystallographers, considered molecular formulae as authentic models of the spatial arrangement of atoms. Starting with typical formulae in which atoms can be substituted, he did not focus on any particular molecule and proposed a general rule valid for any molecule with a fixed geometry:

Let us consider a molecule of a chemical compound having the formula MA^4:
M being a simple or complex radical combined with four atoms A, capable

of replacement by substitution. Let us replace three of them by simple or complex monatomic radicals different from A^1 and also from each other; the body obtained will be unsymmetrical. In other words, the group of radicals RR'R" and A if considered as material points, being each different, forms by itself a structure not superposable on its image ...[13]

Le Bel then considered several molecules with or without rotatory power: for example, the di-substituted derivatives of methane are not optically active, whereas its tri-substituted derivatives are active. From this he derived a general rule on the permanent arrangement of RR'R" radicals around an atom A, in a molecule with rotatory power: 'There must necessarily be a polyvalent atom, around which the radicals affect an asymmetric disposition, and this can only occur if they are not in the same plane.'[14] And he called this atom (whether carbon or nitrogen or some other) asymmetric.

Le Bel therefore concluded that the carbon atom can be asymmetric, i.e. adopt an asymmetric molecular arrangement. However, he treated the carbon atom as an example among others and he was primarily interested in the rotatory power. His aim was to determine *a priori* the rotatory power of chemical compounds in general.

Van't Hoff's reasoning was quite different. Like many of Kekulé's students, he was primarily interested in carbon and its bonds. He tried to reconcile the structural formulae with the reality of the isomers that can be formed by substitution.[15] Wilhelm Körner, who was Kekulé's assistant in Ghent, also worked successfully in this direction. He was able to explain why there are several isomeric derivatives of benzene by the position of the substituent groups in the hexagon: ortho (straight) isomers have the two substituent groups next to each other; meta (beyond) isomers have the two substituent groups separated by a hydrogen; finally, in para isomers, the two substituent groups are in opposite positions on the hexagon (Figure C).

In this way, Körner identified the three isomers of dibromobenzene and proceeded to substitute an NO_2 radical for the remaining hydrogen atoms. With the meta form he obtained three isomers, with the ortho form he obtained two and with the para form only one. By playing with the relative positions of the atoms and substituent groups in this way, Körner was able to predict the existence of some 126 benzene derivatives

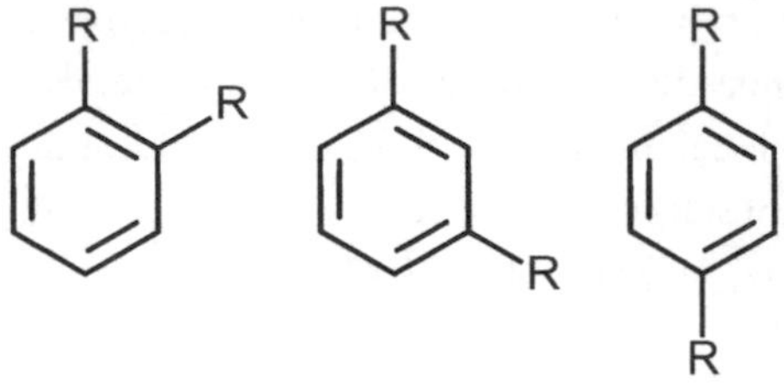

Figure C. Ortho, meta and para positions.

before he synthesized them. Thanks to these assumptions about the relative position of the atoms in the molecule, the chemistry of benzene derivatives became a systematic research programme.

Van't Hoff also tried to predict unknown isomers on the basis of the relative position of the atoms in carbon molecules. He found that the current structural formulae, which represent the atoms bonded to the carbon in the same plane, led to a much higher number of isomers being predicted than those that actually exist, especially in the case of methane derivatives of the type $CH_2(R_1)_2$ (where R stands for the radical, i.e. the bonding possibility).

> [I]f we grant that the affinities of the carbon atom are directed towards the angles of a regular tetrahedron, of which this atom occupies the centre, there results a marked coincidence between the theory and the facts. According to this hypothesis the number of isomers becomes as follows: no isomers in the cases CH_3R_1, $CH_2(R_1)_2$, $CH_2R_1R_2$ and $CHR_2(R_1)_2$; the case alone $CHR_1R_2R_3$, or more generally $CR_1R_2R_3R_4$ allows the prediction of isomerism; in other words, if the four affinities of the carbon atom are saturated by four different groups, we obtain two tetrahedrons of which one is the nonsuperposable image of the other, that is to say, we have to deal with two different structural formulae in space.[16]

In this way, van't Hoff constructed a picture of carbon molecules in space from compounds that have an experimental existence. As much as Le Bel deduced the asymmetric carbon from a general formula, van't Hoff induced a three-dimensional molecular formula from experience. Only by representing the four atomicities in space is it possible to reduce the number of possible isomers to fit the experimental results, reducing the possible to the existing. Van't Hoff then shows that all optically active compounds have an asymmetric carbon. He thus explains the surprising isomerism of maleic and fumaric acids (Figure D).

These bold hypotheses show the extent to which these chemists did not shy away from speculation in order to 'tame' carbon, to find their way through the labyrinth of its compounds and derivatives and to predict their properties. Such boldness aroused the sarcasm and indignation of the German chemist Hermann Kolbe, who was fiercely hostile to atomic theory:

> It is characteristic of the present time with its poverty in and hatred for criticism, that two chemists almost unknown, one from a veterinary college and the other from an agricultural institute, should on the deepest and possibly insoluble problems of chemistry, especially the question of the position of the atoms in space, give their opinions with an amount of assurance and undertake the solution of these questions with an amount of audacity such as to astound the true investigator.[17]

But the asymmetric carbon theory was soon adopted by most chemists.[18] It entered the textbooks in the form that van't Hoff had given it, i.e. as the feature of carbon atoms such that they have the ability to orient their bonds in space. Carbon thus gave rise to 'chemistry in space', to use the catchy title of the brochure published by van't Hoff in 1887. It was the impetus for a new chemistry that Victor Meyer called 'stereochemistry'.

Is molecular asymmetry specific to carbon or not? The question is of critical importance insofar as Pasteur established molecular asymmetry as a dividing line between living products and synthetic products:

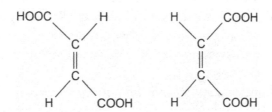

Figure D. Formulae for fumaric and maleic acids.

When the ray of sunlight strikes a green leaf and the carbon of the carbonic acid, the hydrogen of the water, the nitrogen of the ammonia and the oxygen of this carbonic acid and water form chemical compounds and the plant grows, dissymmetric bodies are born. You, on the other hand, however skilful chemists you may be, when you unite these same elements in a thousand different ways, you always make products devoid of molecular dissymmetry. To my knowledge, there is not a single product of chemical synthesis, born under the influence of the causes that can be considered proper to plant life, which is not dissymmetric, which does not have, in other words, the general shape of a helix, of a revolving staircase, of an irregular tetrahedron, of a hand, of an eye. . . . In contrast, there is not a single synthetic product, prepared in laboratories or in mineral nature, which is not in the shape of an octahedron, a straight staircase . . .[19]

It is true that for Pasteur the dissymmetry of molecules is not specific to atoms; it results from the dissymmetry of the *forces* at work.[20] But the question of whether carbon is one of a kind remains unanswered. Could other elements give rise to special chemistries as well? Some chemists did attempt to develop an organic chemistry based on silicon at the beginning of the twentieth century.[21] Frederic S. Kipping in Nottingham devoted his work to the 'organic compounds' of silicon and pointed out the structural analogies with those of carbon.[22] Geoffrey Martin, his collaborator for several years, believed that silicon may have given rise to a form of life that has now disappeared.[23] This hypothesis has now been abandoned, as silicon requires too much energy to break its bonds and authorize the countless reactions that are essential for metabolism. Nevertheless, although silicon was rejected as a rival to carbon in the chemistry of living organisms, it has since taken its revenge in the electronics industry.

However, in 1912, Alfred Werner showed that carbon is not uniquely asymmetric, and that chirality can be observed around a metal atom.[24] Organometallic complexes suggest that the capacity for optical activity is not specific to the carbon atom: the rotatory power of a compound does not depend on the asymmetry of the carbon alone but on all the groups in the asymmetric molecule.

Boron, although unstable and difficult to prepare and preserve, is another candidate on which the German chemist Alfred Stock had long hoped to base a chemistry as rich as that of carbon.[25] Between 1912 and

1937, he successively isolated and systematically studied a host of boron and hydrogen compounds: the tetraborane B_4H_{10}, the diborane B_2H_6, the decaborane $B_{10}H_{14}$, then the pentaborane B_5H_9 or B_5H_{11} and the hexaborane B_6H_{10}. Stock first thought that boron was tetravalent like carbon, but the efforts of several chemists to determine the nature of the boron bonds and the structure of the diborane led to a long debate that did not confirm the similarity to carbon. And even though, during the Second World War, the carbon analogy inspired efforts to produce alternative boron-based fuels, these cost a lot of money and some lives because of the instability of the element's compounds. Carbon thus has no rival.

Dispositions and affordances

Whether or not they are unique to the carbon atom, the ability to form bonds with identical atoms and the ability to form asymmetric molecules are remarkable dispositions.

The carbon–carbon bond is a *disposition* in the precise sense that Aristotle gives to this term (*hexis*, in Greek): it is a way of being (a *habitus* in Latin) characteristic of a thing that manifests itself under certain conditions. For example, being soluble in water is a disposition of sugar in the sense that it manifests itself under a condition which is the presence of water. Similarly, the carbon atom binds to itself under certain conditions, when the atoms with which it binds do not saturate its four valencies. They 'accommodate each other in some manner', as Wurtz said in the passage quoted above. In more recent terms, it is a modal arrangement because the internal characteristics of the carbon atom depend largely on the environment that surrounds it.[26] Whether trivalent or quadrivalent carbon, these are two modes of existence that Gaston Bachelard had clearly pointed out:

> In their substantive form, the two expressions, carbon is quadrivalent and carbon is divalent, are contradictory. In their modal form, once the conditions of deployment of the values in the composition are indicated, both forms are valid and contribute to the deep explanation of the phenomena.[27]

As for asymmetry, it is a disposition in the double sense of the term: an arrangement, a distribution in space and a capacity to do something.

Carbon chemistry seems, in fact, less oriented towards the objective of representing the structure of its compounds by a faithful image than towards understanding what atoms and molecules *do*, what it is in their nature to do, i.e. their capacities. With a firm orientation towards synthesis, it aims less to represent an underlying reality than to attest to a *possibility*. It seeks to delimit the field of possibilities, to play with these possibilities. In short, it is a question of intervening rather than representing, while assuming the reality of what carbon atoms do, as an operational and relational reality.[28]

However, the concept of disposition, insofar as it expresses a permanent capacity that manifests itself under well-determined conditions according to a law, seems to us less adequate when it comes to designating the movement of genesis or instauration of the possibilities opened up by the carbon atom – that is to say, the work of synthesis itself rather than the stabilized properties of the products of synthesis.[29] Indeed, the 'game of possibilities' on which carbon chemists embark is not like that of Darwinian evolution: a blind game, without a plan. Chemists seize the opportunities offered by the C–C bond or by asymmetry in order to create molecules of interest – cognitive, industrial, pharmaceutical, military or other, depending on the context. The English term *affordance* therefore seems more appropriate. Introduced by the psychologist James Gibson as part of his ecological theory of visual perception, this term refers to what a thing or an environment offers in connection with an action.[30] Our world is made up of objects that invite a particular action. The concept of affordance expresses this idea perfectly: objects offer possibilities for action that depend on the context and the physiological and intentional configuration of a subject. For example, surface water offers an animal of the plains the possibility of cooling off and quenching its thirst. Frozen, it invites skating in other contexts and for other subjects. A stone offers a shelter to a crawling animal, the possibility of a hidden prey for a cat or the possibility of a tool for a human. Affordance is neither a real, objective property of things that we measure in physical terms, nor a subjective quality that we would attribute to them from a mental image.

> The concept of affordance is derived from these concepts of valence, invitation, and demand but with a crucial difference. The affordance of

something does not change as the need of the observer changes. The observer may or may not perceive or attend to the affordance, according to his needs, but the affordance, being invariant, is always there to be perceived. An affordance is not bestowed upon an object by a need of an observer and his act of perceiving it. The object offers what it does because it is what it is. To be sure, we define what it is in terms of ecological physics instead of physical physics, and it therefore possesses meaning and value to begin with. But this is meaning and value of a new sort.[31]

Crossing the dichotomies between objective and subjective, between physical and psychic, the concept of affordance establishes a bridge, a relationship between a being and its environment. It is precisely this relationship that is appropriate to carbon. It prevents us from confining it to a specific class of objects such as chemical element, material, nuisance or pollution. The carbon atom offers multiple possibilities of action in relation to specific situations. These constitute the identity of carbon as much as that of the living things that benefit from it for their survival or their career. How, then, can we understand why chemists persist in talking about 'organic chemistry' (and inorganic chemistry)? Certainly, carbon chemistry has an obvious connection with the biochemistry of organized beings, but organic molecules can be produced by processes unrelated to living organisms, and living organisms also depend on inorganic chemistry (enzymes, for instance, often include transition metals like iron and copper, not to mention bones, teeth and shells). After almost two centuries of prolific production of both knowledge and species of carbon compounds, this chemistry we persist in calling 'organic' has not brought us decisively closer to the threshold of the organic processes of living things. It has managed to synthetically (re)produce a number of molecules derived from living organisms – from urea to vitamin C – but by no means the formative processes of the living organisms themselves. Once a weapon against vitalism, the term 'organic chemistry' seems more like a misnomer today, where no one subscribes to the idea of a 'vital force'. So why not simply make carbon the singular hero of an entire branch of chemistry? After all, it is the bonding patterns of carbon that bring some order to the myriad of organic compounds through the categories of aromatic and aliphatic, cyclic and acyclic or saturated and unsaturated. Carbon offers itself as a guide in the labyrinth of countless molecules, a labyrinth that extends over both the living and the non-living.

Our guess is that the term 'organic chemistry' is resistant precisely because carbon is presented as a bonding entity, as a *connector* rather than as a *substance* defined by its essence. Its individuality is signalled less by its intrinsic properties than by the multiplicity of associations it can form by playing on the carbon–carbon bond and the asymmetry that widens the field of possibilities.

A philosopher's stone

With its mythical aura, the philosopher's stone is a good metaphor for the carbon–carbon bond in the sense that it opens doors, offers a path of action and reveals a horizon of possibilities. It deploys affordances. But unlike the philosopher's stone, which only revealed its potential after a long initiation by alchemists, carbon molecules themselves inspire chemists with paths of transformation. In this respect, they present a typical case of affordance in the sense of the capacity of a thing to suggest its own use. To conclude, let us cite three examples of the exploitation of the bonding potential of carbon: the benzene cycle, macromolecules and metathesis.

The carbon–carbon bond has made the chemical industry wealthy by opening the way for so-called 'aromatic compounds'. Extracting all sorts of colours – mauve, fuchsia, magenta, indigo – from black coal tar, an undesirable by-product of coke production, was possible thanks to the stability of the ring formed by the six carbon atoms. A hydrogen atom can actually be substituted for an atom of another element or a functional group in order to obtain a derivative which, in turn, can serve as a basis for synthesizing the molecules that are of interest. The first generation of synthetic dyes based on aniline was synthesized by William Henry Perkin in the 1850s. This paved the way for a second generation of synthetic dyes based on alizarin, which displaced plant-based dyes for good towards the end of the nineteenth century. Since then, benzene derivatives have formed the range of basic products – nitrobenzene, chlorobenzene, aniline, phenol, etc. – that have made the fine chemicals industry rich for the past century and a half.[32] Medicines, solvents and paints use significant quantities of benzene.[33] Global production reached about fifty million tonnes in 2012, although various uses of benzene are severely restricted because of its harmfulness to the environment and health. Like carbon dioxide, the carbon–carbon bond offers poisons and cures: the beautiful flat hexagonal structure of benzene is capable of intercalating between the bases of DNA and initiating cancers as well as producing therapeutic compounds.

In 1926, Hermann Staudinger said: 'We are at the very beginning of the chemistry of real organic compounds and we are far from seeing its conclusion.'[34] Why did this Freiburg chemist, who was interested in synthetic rubber, envisage a new chemistry with 'real organic compounds'? This was during a controversy about the nature of polymers, i.e. compounds consisting of several (*poly*) parts (*meros*), such as olefins (C_nH2_n), which all have the same composition but different molecular weights. Some chemists saw them as colloids, i.e. aggregates of molecules bound by physical forces, whereas Staudinger considered them to be macromolecules.[35] These giant molecules, which can contain from a thousand to a hundred thousand atoms, are formed by the covalent linking of units such as CH_2. By calling for the integration of polymers into Kekulé's organic chemistry, Staudinger revealed a new facet of the bonding potential of carbon atoms.

What Staudinger revealed when he invented the concept of the macromolecule in 1922 was that the bonding potential of carbon extends beyond the formation of rings. The covalent linking of carbon units forms kinds of necklaces, or linear chains, onto which side chains can be attached. The bonding potential mobilized to form macromolecules is deployed between carbon and carbon (as in polyethylene or polypropylene) or between carbon and oxygen (as in the case of polyethers or polyesters) or between carbon and nitrogen (in the case of polyamides). These polymers are formed from liquid solutions, more or less viscous depending on their degree of polymerization. They have enormous molecular masses compared to monomers, so that their properties derive not only from the arrangement of the atoms in the molecule, but also from conformation, i.e. the shape that these chains take in space. Covalent bonding between monomers opened up a new field of possibilities that was explored in the 1920s and 1930s in university and industrial laboratories, both in Germany and at DuPont in the United States.[36] This was a world where the natural and the artificial constantly met and enriched each other, because carbonaceous macromolecules constitute the living tissues of the plant and animal worlds, as well as most of the synthetic fibres and materials we use in everyday life.

Finally, just as the philosopher's stone was used as a means of transforming vile bodies into noble ones, the carbon–carbon bond offers the freedom to modify a molecular structure to suit our purposes. Olefins, produced by the distillation and cracking of petroleum, are hydrocarbons with single and double carbon–carbon bonds. It is now accepted that in these double bonds, two pairs of electrons circulate and are exchanged. This is why these molecules lend themselves to a kind of dance involving the swapping of partners that shifts carbon patterns (Figure E).

The exchange is done with the help of a transition metal, which makes a temporary bond with one of the carbon units and allows the carbon–carbon bonds to be shifted. This phenomenon is called 'metathesis' (the Greek for 'permutation') and offers an elegant and rapid method of synthesis. It was first observed, then described and implemented in industrial laboratories, at DuPont in the 1950s, then at the French Petroleum Institute by Yves Chauvin in the 1960s. Chauvin was awarded the Nobel Prize in Chemistry, shared with Richard Schrock and Robert Grubbs in 2005.[37] Metathesis has many industrial applications: in petrochemicals, where it allows better use of by-products of petroleum cracking; in the pharmaceutical industry; in the synthesis of plastics; and even in biochemistry.

These three examples, at the frontier between the academic and industrial worlds, illustrate the treasure trove of affordances hidden in the bonds of the carbon atom. These bonds are both stable and diversified

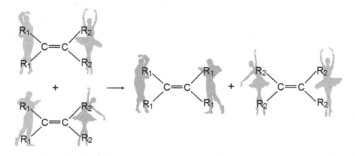

Figure E. Metathesis diagram.

and are exploited by nature and industry alike. Thanks to the polysaccharide macromolecules that contain genetic information and the range of synthetic polymers that populate our environment, living things and many technical objects owe their existence to the innumerable bonding capacities offered by carbon.

7

Welcome to the nanoworld

How did the tiny prefix 'nano' in grant applications leverage millions of dollars for scientific research at the nanoscale? This question, raised in 2006 by a non-governmental organization (ETC Group), was ironizing about all the noise and excitement over access to the billionth of a metre scale, i.e. the molecular scale.[1] The major research programmes launched in the 2000s carried slogans such as 'shaping the world atom by atom'[2] or 'there's plenty of room at the bottom'. After the conquest of space, here comes the conquest of the nanoworld. A distant world, populated by objects that were previously inaccessible to our senses, is being transformed into a land of promise, inhabited by atoms, genes, information bits and neurons.

Carbon is the king of this giddy world, while fullerenes, nanotubes and graphene are the stars of nanotechnology (Figure 5). These carbon molecules have the power to attract thousands of researchers and millions of dollars, and to make people dream of miracle solutions to all the problems of electronics, health, the environment, etc. But where do these nanocarbons come from? For centuries, generations of scientists have described the characteristics, properties and behaviour of carbon, and chemists and engineers have exploited its properties and affordances. Could they have missed out on these wonderful molecules? Have they simply been revealed by powerful scientific instruments such as the scanning tunnelling microscope, or have they been forged by skilful molecular architects? In short, are they discoveries or inventions? Carbon certainly has more than one trick up its sleeve; it's always coming up with surprises and writing new stories into the history of humanity.

There is scarcely any substance which acts a more conspicuous part in the economy of nature than carbon. In the animal, the vegetable, and the mineral kingdoms, it is equally abundant and useful, nor can anything be more instructive than to study the infinite variety of purposes which it serves, or

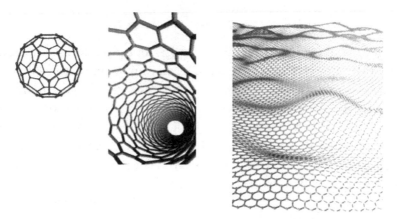

Figure 5. Fullerene, carbon nanotube and graphene.

more interesting than to observe the wonderful diversity of forms which it puts on. What appearances can be in more perfect contrast to each other than those of soot on the one hand and of the diamond on the other? Yet carbon embraces both extremes. … It was a discovery which excited great attention at the time when Lavoisier, and after him Mr. S. Tennant, unveiled carbon from beneath this last disguise; if, indeed, this expression be philosophically admissible, and if it be not impossible to say that any one state of existence is more natural or more proper to the substance than another. For although a form very different from that under which a substance has been commonly known may be popularly termed a *disguise*, under which it has concealed itself, yet it is not more truly so than any of its other shapes; and it is always a step in the progress of science when an entirely new form, especially of an important body, is obtained. This is in fact to approximate the laboratory of the chemist to the great officina of nature, where many a substance may exist in states very different indeed from any which the industry of the experimenter has enabled him to discover, or his art to make it assume.[3]

These reflections on the disguises of carbon could be applied to fullerenes, nanotubes or graphene, if the somewhat convoluted phrasing of the quotation did not betray a past style. They actually date from 1826 and were inspired in a certain Dr Hugh Colquhoun by the discovery of 'several highly interesting states of aggregation of carbon, one of which is not only a very singular structure, but also an entirely new form'.

Filaments doomed to oblivion

Some general considerations on the 'disguises' of carbon formed the start of an article reporting the observation of an unwelcome carbon deposit in a steel-making furnace. The scene was Crossbasket Castle, Blantyre, near Glasgow. In the absence of the squire, who was none other than the chemist Charles Macintosh – already famous for his invention of the waterproof fabric that took Britain by storm – Colquhoun was supervising experiments to develop a new steel-making process developed by Macintosh in an experimental factory. The 'Macintosh process' involved reacting molten iron with a hydrocarbon gas in a porcelain vessel at high temperature.[4] When the gas is in excess, a deposit is formed which Colquhoun described as 'threads of carbon', 'mineral hair', 'a single lock ... seems to contain thousands of these thin filaments ... as delicate as the filaments of the lightest spider's web'.[5] To characterize this strange product, he mixed it with various salts and acids, heated these mixtures and found that this deposit was capable of decomposing certain substances by causing deflagrations, as would charcoal or plumbago.[6] He concluded that it was a very stable carbonaceous substance that was almost devoid of hydrogen and oxygen, probably an entirely new form of 'pure metallic carbon'. But the story was short-lived, as Macintosh had less luck this time around than with his raincoat. His steel-making process required very high temperatures and did not reach industrial scale, so it was quickly abandoned.

However, carbon filaments resurfaced in the *Comptes rendus de l'Académie des sciences* towards the end of the nineteenth century, where they were described twice in two different and independent contexts. First, the Alsatian chemist Paul Schützenberger, director of the École municipale de physique et de chimie in Paris, and his son Léon, a chemical engineer who graduated from the same school, were trying to decompose 'azotocarbons' such as cyanogen ($N\equiv C–C\equiv N$).[7] After lining a 'retort carbon pod'[8] with cryolite powder, a powerful solvent, and heating it to a high temperature, they found that after an hour and a half the cyanogen gas was decomposed into carbon and nitrogen. This 'filamentous product . . . capable of agglomerating by pressure and friction into a graphitic mass' was a strange kind of carbon. Its oxidation products did not allow it to be identified with any of the three varieties of graphite already described in the literature. So the Schützenbergers cautiously concluded: 'It follows that the filiform carbon formed by the pyrogenic decomposition of cyanogen in the presence of cryolite vapours constitutes a variety of carbon close to electrical graphite but not identical with it.'[9] However, as this filiform carbon was of no industrial interest in a school and at a time when applied work[10] took precedence, the matter ended there.

A few years later, carbon filaments appeared once again in coke ovens. Two industrialists, Constant and Henri Pélabon, presented a paper to the Academy of Sciences in Paris, read by Henri Moissan,[11] on a curious 'coal wool' formed in the upper part of coke ovens: 'The filaments making up the black wool are dull, their surface is covered with asperities sometimes arranged very regularly; the threads then seem to be formed by a succession of rings.' Their thickness was about one to fifteen microns and their length varied between five and eight centimetres. This thread-like carbon was an undesirable curiosity because its presence was a sign that the furnaces had been pushed to too high a temperature. The authors quoted the Schützenberger paper but surprisingly neither of them referred to the carbon filaments that Thomas Alva Edison used with relative success in his incandescent light bulbs.

A startling omission, because Edison was a man who knew how to make a name for himself. His lamps were shining at the Paris World Fair in 1889. Indeed, carbon filaments were already widely known to the public, as electric light was the star of the celebrations of technical progress at such World's Fairs.[12] Edison chose carbon to create incandescence in electric lamps after a vast research programme in which he spent forty thousand dollars and tested six thousand natural substances.[13] In 1879, he filed a patent describing the processes for obtaining carbon filaments, made up of twists of several wires, each about ten microns in diameter.

The first incandescent lamps, marketed in the 1880s by the Ediswan firm (a contraction of Thomas Edison and Joseph Swan) for street lighting, contained 'metallized carbon' filaments obtained by 'flash carbonization'[14] of bamboo fibres, which are very rich in cellulose (Figure 6). However, the carbon filament was soon supplanted by tungsten because overheated carbon sublimates causing the vapour to condense on the glass envelope, blackening the bulb after thirty hours

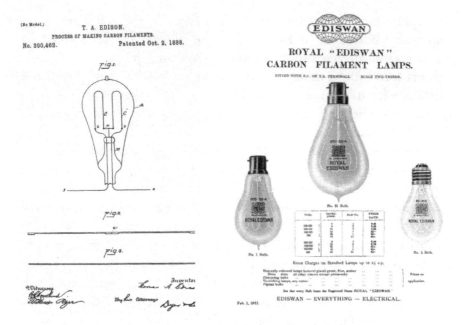

Figure 6. Carbon filament lamps, Edison patent, 1888; Ediswan catalogue, 1921.

of use. So carbon filaments abandoned the world of lighting . . . even if some nanotechnology enthusiasts are now betting on a comeback for luminescent carbon with the aid of multi-walled nanotubes.[15]

Filamentary carbon thus made quite a few appearances at the forefront of technical innovation in the early twentieth century. It attracted attention several times, but had only fleeting successes before being called back to the stage in the 1950s thanks to new instruments such as the electron microscope.

For instance, some strange 'carbon vermicules' measuring from a hundred angstroms to two microns suddenly popped up under the electron microscope of three British chemists, W. R. Davis, R. J. Slawson and G. R. Rigby.[16] These vermicules were so strong that they penetrated the linings of blast furnaces and contributed to their wear and tear, even breaking the bricks. They were therefore undesirable. The paper that Davis et al. wrote on this 'unusual form of carbon' conjectured that their growth was catalysed by metal particles embedded in the brick, which was confirmed by further observations by two Soviet researchers who published electron microscope images of iron-catalysed carbon filaments with an internal cavity of about fifty nanometres.[17] In retrospect, these images look like multi-sheet carbon nanotubes. But the paper, published in Russian, did not attract the attention of the broader scientific community.

During the 1950s, these strange filaments made a new foray into scientific news thanks to work on graphite, a strategic material for nuclear reactors at that time.[18] Two crystallographers, Mats Hillert and Nils Lange, who were interested in the 'filamentary growth of graphite', directed a whole arsenal of instruments onto these filaments: polarized light microscopes, electron microscopes, X-ray diffraction and electron diffraction.[19] Hillert and Lange noted that metallic particles clung to these filaments; they pointed out that they were hollow, formed by layers of lamellae wound into cylinders, with one or more walls, and that they could take on several shapes through growth, while retaining the same crystalline structure, that of graphite.

Once again, it is possible to read this 1958 article as an early description of carbon nanotubes. Some even claim that nanotechnology is merely the rediscovery of previously observed phenomena that have been forgotten.[20] Wasn't it established as early as the 1950s that a nanotube could be produced by catalytic growth?[21] Undoubtedly this was written about, but how much was known about it? In reality, these retrospective readings shed less light on obscure forgotten precursors likely to undermine the revolutionary claims of nanotechnologies than on screening effects produced by research programmes. Indeed, in the 1950s, the catalytic growth of carbon filaments could have become a field of research in its own right, if these filaments had not been eclipsed by carbon fibres.

Seeing without discovering

In 1957, American physicist Roger Bacon observed 'graphite whiskers' obtained during a study of the melting point of graphite under high pressure and temperature as hydrocarbons were vaporized by electric arc discharge in a gas.[22] He carefully described how these whiskers appeared when the pressure was reduced, and formed a kind of stalagmite through direct passage from gas to solid phases. When he broke and opened one of these whiskers to see its structure, he observed concentric tubes formed by a winding layer of graphite.[23]

A few decades later, Bacon frankly confessed: 'I may have made nanotubes but I didn't discover them.'[24] That was very well put. One can indeed observe without discovering. Archimedes' famous eureka moment is less a matter of evidence than a disposition of mind. Carbon nanotubes did not come into existence thanks to the visual evidence provided by powerful instruments such as electron or, later on, scanning tunnelling microscopes. They were also the result of an expectation created by a scientific, political, economic and cultural context. Unlike the chemists who observed the filaments produced in the blast furnaces to try to get rid of them, Bacon, a researcher at Union Carbide, a large chemical company, saw in these filaments possible precursors of carbon fibres to reinforce plastics more effectively than glass fibres. He saw a future for carbon in the age of plastics.[25] Initially made from coal or petroleum pitches, carbon fibres were now produced industrially from PAN (polyacrylonitrile $[C_3H_3N]_n$). As the manufacturing process, requiring carbonization and graphitization, was rather complex and costly, intensive research into the manufacture of carbon fibres was being carried out worldwide.

Thus, in the 1970s, researchers' attention shifted from the fibre itself to the fibre formation process with the hope of replacing PAN by cheaper methods starting from raw, primary materials. Morinobu Endo, a young Japanese chemical engineer, wanted to get into carbon fibres. In 1974, he travelled in France to Orléans to do research in Agnès Oberlin's laboratory. There they studied a process for growing carbon filaments in the vapour phase by catalytic decomposition of benzene developed in the laboratory. During the formation of fibres in the vapour phase, Endo again observed a 'hollow tube', which he also called 'central tube'.[26] Twelve years later, he suggested that this hollow tube was not a by-product of fibre manufacture, but an essential initial step in fibre formation. This takes place in two stages: first, catalytic growth of a hollow tube, then thickening by pyrolytic carbon deposition.[27] The hollow structure was thus no longer a defect that needed to be removed, but the heart of the fibre, and perhaps the cause of its remarkable mechanical properties. Unlike Bacon, Endo demonstrated that carbon fibre is primarily a tube. In a short time, he too opened up a successful path for these carbon tubes in industrial applications.[28] Had he discovered carbon nanotubes? It takes more than that to bring them into existence . . . There is a missing eureka moment, which did appear, but where it was least expected, in the course of a study of interstellar space.

A football

Harold Kroto, an astrochemist at the University of Sussex in the UK, had been working on mysterious 'diffuse interstellar bands' (DIBs) detected by radio frequency in 'dark nebulae', regions where dust from the interstellar medium seems to concentrate in large dark clouds. A study with the Algonquin Observatory (Ontario) radio telescope determined that these bands were the signatures of long molecules of the polynylcyanides HC_5N (H–C≡C–C≡C–C≡N), HC_7N (H–C≡C–C≡C–C≡C–C≡N), and HC_9N, 'the largest molecule detected in interstellar space'.[29] Kroto was guessing that these molecules originated in red giant stars, and to test this hypothesis he travelled to the laboratory of chemist Richard Smalley at Rice University in Texas, which had a homemade device that, Kroto believed, would allow the simulation of their formation by reproducing the extreme conditions of such an environment. As for Smalley, he had much more mundane intentions, related to the production of nanometric clusters (at first metallic, and then – why not? – carbon-based), of interest for microelectronics and the semiconductor industry.[30]

Smalley's nameless machine, referred to as the AP2 for 'second-generation apparatus', consisted of a supersonic nozzle, producing a very high-energy laser beam that struck a rotating graphite disc. On the surface of the irradiated graphite, a hot and chaotic plasma was produced: carbon atoms vaporized into a dense and rapid flow of helium. Using a mass spectrometer, Kroto and Smalley detected a few peaks in this flow of helium that indicated the presence of 'something' bizarre but very stable. These were carbon clusters, consisting of sixty atoms. Kroto, Smalley and colleagues announced these newcomers in the journal *Nature* in 1985 (Figure 7).

While every chemistry paper must offer a structural formula with a graph of the molecule described, Kroto and his collaborators were content with a photograph of a football (soccer ball in American) on a Texas lawn.[31] How could a metaphor put forward on the basis of a single experiment convince the editors and referees of *Nature*, who are known to be tough? 'It was such a beautiful and perfect structure, how could

162   LETTERS TO NATURE   NATURE VOL. 318 14 NOVEMBER 1985

C_{60}: Buckminsterfullerene

H. W. Kroto[*], J. R. Heath, S. C. O'Brien, R. F. Curl
& R. E. Smalley

Rice Quantum Institute and Departments of Chemistry and Electrical
Engineering, Rice University, Houston, Texas 77251, USA

Fig. 1 A football (in the United States, a soccerball) on Texas grass. The C_{60} molecule featured in this letter is suggested to have the truncated icosahedral structure formed by replacing each vertex on the seams of such a ball by a carbon atom.

During experiments aimed at understanding the mechanisms by which long-chain carbon molecules are formed in interstellar space and circumstellar shells[1], graphite has been vaporized by laser irradiation, producing a remarkably stable cluster consisting of 60 carbon atoms. Concerning the question of what kind of 60-carbon atom structure might give rise to a superstable species, we suggest a truncated icosahedron, a polygon with 60 vertices and 32 faces, 12 of which are pentagonal and 20 hexagonal. This object is commonly encountered as the football shown in Fig. 1. The C_{60} molecule which results when a carbon atom is placed at each vertex of this structure has all valences satisfied by two single bonds and one double bond, has many resonance structures, and appears to be aromatic.

The technique used to produce and detect this unusual molecule involves the vaporization of carbon species from the surface of a solid disk of graphite into a high-density helium flow, using a focused pulsed laser. The vaporization laser was the second harmonic of Q-switched Nd:YAG producing pulse energies of ~30 mJ. The resulting carbon clusters were expanded in a supersonic molecular beam, photoionized using an excimer laser, and detected by time-of-flight mass spectrometry. The vaporization chamber is shown in Fig. 2. In the experiment the pulsed valve was opened first and then the vaporization laser was fired after a precisely controlled delay. Carbon species were vaporized into the helium stream, cooled and partially equilibrated in the expansion, and travelled in the resulting molecular beam to the ionization region. The clusters were ionized by direct one-photon excitation with a carefully synchronized excimer laser pulse. The apparatus has been fully described previously[2-5].

The vaporization of carbon has been studied previously in a very similar apparatus[6]. In that work clusters of up to 190 carbon atoms were observed and it was noted that for clusters of more than 40 atoms, only those containing an even number of atoms were observed. In the mass spectra displayed in ref. 6, the C_{60} peak is the largest for cluster sizes of >40 atoms, but it is not completely dominant. We have recently re-examined this system and found that under certain clustering conditions the C_{60} peak can be made about 40 times larger than neighbouring clusters.

Figure 3 shows a series of cluster distributions resulting from variations in the vaporization conditions evolving from a cluster distribution similar to that observed in ref. 3, to one in which C_{60} is totally dominant. In Fig. 3c, where the firing of the vaporization laser was delayed until most of the He pulse had passed, a roughly gaussian distribution of large, even-numbered clusters with 38-120 atoms resulted. The C_{60} peak was largest but not dominant. In Fig. 3b, the vaporization laser was fired at the time of maximum helium density; the C_{60} peak grew into a feature perhaps five times stronger than its neighbours, with the exception of C_{70}. In Fig. 3a, the conditions were similar to those in Fig. 3b but in addition the integrating cup depicted in Fig. 2 was added to increase the time between vaporization and expansion. The resulting cluster distribution is completely dominated by C_{60}, in fact more than 50% of the total large cluster abundance is accounted for by C_{60}; the C_{70} peak has diminished in relative intensity compared with C_{60}, but remains rather prominent, accounting for ~5% of the large cluster population.

Our rationalization of these results is that in the laser vaporization, fragments are torn from the surface as pieces of the planar

graphite fused six-membered ring structure. We believe that the distribution in Fig. 3c is fairly representative of the nascent distribution of larger ring fragments. When these hot ring clusters are left in contact with high-density helium, the clusters equilibrate by two- and three-body collisions towards the most stable species, which appears to be a unique cluster containing 60 atoms.

When one thinks in terms of the many fused-ring isomers with unsatisfied valences at the edges that would naturally arise from a graphite fragmentation, this result seems impossible: there is not much to choose between such isomers in terms of stability. If one tries to shift to a tetrahedral diamond structure, the entire surface of the cluster will be covered with unsatisfied valences. Thus a search was made for some other plausible structure which would satisfy all sp^2 valences. Only a spheroidal structure appears likely to satisfy this criterion, and thus Buckminster Fuller's studies were consulted (see, for example, ref. 7). An unusually beautiful (and probably unique) choice is the truncated icosahedron depicted in Fig. 1. As mentioned above, all valences are satisfied with this structure, and the molecule appears to be aromatic. The structure has the symmetry of the icosahedral group. The inner and outer surfaces are covered with a sea of π electrons. The diameter of this C_{60} molecule is ~7 Å, providing an inner cavity which appears to be capable of holding a variety of atoms[8].

Assuming that our somewhat speculative structure is correct, there are a number of important ramifications arising from the existence of such a species. Because of its stability when formed under the most violent conditions, it may be widely distributed in the Universe. For example, it may be a major constituent of circumstellar shells with high carbon content. It is a feasible constituent of interstellar dust and a possible major site for

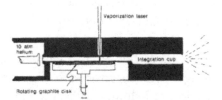

Fig. 2 Schematic diagram of the pulsed supersonic nozzle used to generate carbon cluster beams. The integrating cup can be removed at the indicated line. The vaporization laser beam (30–40 mJ at 532 nm in a 5-ns pulse) is focused through the nozzle, striking a graphite disk which is rotated slowly to produce a smooth vaporization surface. This helium carrier gas provides the thermalizing collisions necessary to cool, react and cluster the species in the vaporized graphite plasma, and the wind necessary to carry the cluster products through the remainder of the nozzle. Free expansion of this cluster-laden gas at the end of the nozzle forms a supersonic beam which is probed 1.3 m downstream with a time-of-flight mass spectrometer.

* Permanent address: School of Chemistry and Molecular Sciences, University of Sussex, Brighton BN1 9QJ, UK.

Figure 7. H. W. Kroto, J. R. Heath, S. C. O'Brien, R. F. Curl and R. E. Smalley, 'C_{60}: buckminsterfullerene' © Macmillan Publishers Ltd., *Nature*, 318, pp. 162–3, doi: 10.1038/318162a0, p. 162, 1985.

it be wrong?' Kroto declared a few years later.[32] It became real because a structure so mathematically perfect as a truncated icosahedron could not fail to exist.

In fact, as Kroto and Smalley learned only after the release of the *Nature* report, spheric 'hollow molecules' of carbon had already been predicted as early as 1966 by a fictional inventor known as Daedalus (David Jones) who ran a speculative science column in *The New Scientist*, 'The inventions of Daedalus'.[33] In addition to Daedalus' speculations, C_{60} and bigger polyhedral carbon clusters had been repeatedly postulated and subjected to theoretical calculations. Japanese computational chemist Eiji Osawa predicted that in such a structure carbon would be 'superaromatic' (i.e. a conjugated aromatic structure that goes on and on and wraps back around itself) and would thus probably be stable.[34] He displayed a football image on the front page of the publication. Several other calculations followed,[35] including a paper entitled 'Footballene: A theoretical prediction for the table, truncated icosahedral molecule C_{60}'.[36] None of these theoreticians cited one another. Theoretically possible and stable, the carbon ball also represented a synthetic challenge for generations of chemists trained in 'futile attempts to synthesize "socchorene".'[37]

Renamed fullerene by analogy with the domes designed by the futuristic architect Buckminster Fuller, the molecule was striking in its beauty. An aerial beauty, however, because these fullerenes were minuscule traces detected in a flow of helium, 'a puff in a helium wind'.[38] They still only exist as laboratory curiosities. However, fullerenes became a worldly reality five years later when Wolfgang Krätschmer and his team succeeded in producing a macroscopic crystal of these solid footballs, known as fullerite, using an electric arc.[39] These molecules are very pure carbon consisting of 90% C_{60} and 10% C_{70}. Here, then, is a new allotrope of carbon. In fact, this molecule that Kroto was chasing in deep interstellar space has been around on earth for a long time in chimney soot and lightning strike sites. But it has only really existed and been in the news since we learned how to produce it in bulk quantities using Krätschmer's simple and replicable method. Since then, the astrochemical question – whether it is fullerenes, if not polynylcyanides, that fit the DIBs spectra – has faded to the background. Today most fullerene researchers do not care about it – staging the molecules floating like celestial spheres in far-out space provides them simply with a nice aura of mystery. Although Kroto kept working on the problem, the mundane object itself – fullerene – has far overtaken the scientific question that initially prompted its discovery. It has multiplied and created a new world, the world of nanocarbons.

The nanotube jungle

The discovery of carbon nanotubes is usually traced back to a 1991 paper by Sumio Iijima, an electron microscopy researcher at the Japanese micro-electronics firm NEC.[40] Cited more than fifty thousand times, this brief article entitled 'Helical microtubules of graphitic carbon' is seen as the dawn of a new era, a first step into the nanoworld. Why this paper and not one of the many others that had already observed and described such tubes many years before?[41] One cannot even invoke the magical power of the prefix 'nano' to justify the fame of this article, since it speaks of 'microtubules' or 'needle-like tubes' (the term 'nanotube' appeared only slightly after[42]). Moreover, and unlike many nanotech-related papers, the article does not promise a bright future of industrial applications, it simply describes tubes obtained by the electric arc method. In any case, the vapour-grown multi-walled nanotubes made by Endo were already mass-produced for electric batteries for more than a decade when he learned – much to his stupefaction – that they had been 'discovered' by his Japanese colleague.[43] And even Iijima himself had already discovered nanotubes in 1980 in two papers describing a set of clusters of carbon sheets obtained by vapour deposition of amorphous carbon films.[44] So why 1991 as the date of discovery?

The first answer is an experimental one. It lies in the precise location where Iijima collected these tubes. He looked for and found them not in the soot deposits left by the vaporization of carbon, but on the graphite electrodes that were used to produce the electric discharge in the Krätschmer methods. In other words, he found them not in the *products* generated by the experimental system but on the *apparatus of production* itself. In a way, Iijima's paper shifted the focus and seems to say, 'Hey, guys, look over here and not over there.' Since the experimental set-up was quite simple, the technique for producing nanotubes became accessible to any researcher.

A second, more subtle, reason lies in the dual process of decontextual-ization and recontextualization that made these carbon tubes, which had been described for decades, significantly more interesting and attractive. Iijima's short article presented the tubes as objects that were made from graphite alone, without catalysts or hydrocarbons. They were thus cut off from the industrial world in which they were originally noticed, totally detached from coal, blast furnaces or reinforcing fibres for composites. Having cleared the air around them, they appeared new, unique, as if coming from another, or not-yet-existing, world. As for recontextual-ization, Iijima referred directly to fullerenes and to Krätschmer's paper. He suggested that nanotubes could be closed at their ends by half-spheres. They were thus 'finished', brought to completion, 'capped' as it were, with half-fullerenes. Fullerenes could be seen as cut carbon tubes, and carbon tubes as elongated fullerenes. The paper also explained, with great emphasis on geometric diagrams, the formation of both carbon tubes and fullerenes 'by their projection on a honeycomb lattice' according to a certain winding angle that determines their kind of growth (spherical or tubular, scrolled or spiralled), geometry ('armchair' versus 'zigzag' or 'chiral') and helicity ('helical pitch'). In this way, these tubes were now included in a large family of hollow objects (Figure 8). A family that was constantly growing, with single- or multi-walled nanotubes, nanotubes

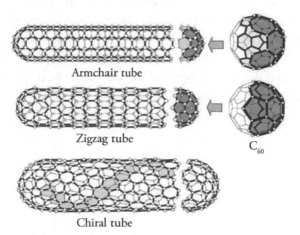

Figure 8. Various nanotube structures obtained from different hemisphere cuts of a fullerene (M. S. Dresselhaus, G. Dresselhaus and R. Saito, 'Physics of carbon nanotubes', *Carbon*, 33/7, 1995, pp. 883–91).

of all diameters and shapes: onion-shaped, sea urchin-shaped, necklace-shaped, serpentine, helix-shaped, spiral-shaped, tripod-shaped, etc.[45] These products of the electric arc fired researchers' imaginations.

These small objects that could be considered mere curiosities were, so to speak, helping and supporting each other as they came into existence. This 'inter-objective' relationship was much more important than the subject–object relationship that could be established between Iijima and nanotubes, between discoverers and *their* discovery. Nanotubes and fullerenes are henceforth a class of objects, an exotic family that keeps on producing hollow and prodigious beings full of promise. The hollow itself, in fact, becomes another affordance. Whereas Endo's hollow central tube arranges a fibre to exhibit interesting mechanical properties, Iijima suggested that due to their helical configuration, tubes can offer special electronic functions: depending on their winding angle, length, curvature, diameter and number of walls, they are either conductive or semiconductive, and their electronic behaviour can be accurately predicted from and shaped by their geometry.[46]

It is definitely impossible to assign a precise date for the discovery of nanotubes. Nor can it be attributed to an individual or even an instrument that revealed them. Rather, this chapter has suggested that carbon has repeatedly tried to attract the attention of researchers by 'disguising' itself as filaments, whiskers, tubes and balls. It has come in many shapes, but their arrangements and affordances remained buried in the black soot of furnaces. Nanotubes only really came into existence in association with fullerenes as the heads of a large family of world-making nano-objects. They opened the doors to the nanoworld that has captivated thousands of researchers.

8

Strategic materials

In May 2011, the European Commission together with national governments and industry made a grant of one billion euros over ten years for graphene research. The 'Graphene Flagship' brought together hundreds of European laboratories working on the structure, properties and performance of graphene and its possible applications in the fields of electronics, energy and medicine.[1] Europe was not alone in the race: the United States, China, Japan, Korea and Singapore were also banking on this 'strange' carbon.

So what was it about this allotrope of carbon that caused such a massive mobilization? It was, after all, only a thin black film obtained by two researchers from the University of Manchester, Andre Geim and Konstantin Novoselov, by peeling adhesive tape off a piece of graphite, yet this work earned them the Nobel Prize in Physics in 2010.

The contrast between the banality of such an object and the marvellous feats expected of it is disturbing. Graphite, of course, had been a commonplace object for centuries, particularly in its use as pencil lead.[2] How could it suddenly, at the turn of the twenty-first century, reveal totally new capacities and behaviours? How could carbon, so ordinary, familiar, well known and long mastered, suddenly become so strategic that states were banking on it to consolidate their military, economic and political power?

Nuclear graphite

This was not the first time that graphite had attracted the attention of high-tech innovators. Indeed, in the mid-twentieth century, graphite became a highly strategic material in the competition for nuclear technologies.[3] But what does it mean for a well-known carbon allotrope like graphite to become a nuclear material?

The alliance between graphite and nuclear energy is based on one question: how does one produce a chain reaction? To do this, the neutrons emitted during the fission of uranium atoms must be slowed down with the help of a liquid or solid moderator, which absorbs part of the neutrons' energy. There are three candidates that have a mass close to that of the neutrons. The deuterium contained in heavy water was the first candidate chosen by Frédéric Joliot on the eve of the Second World War, but this held up his work for a time as producing heavy water is complicated and expensive. Beryllium would be another option, but it was also rejected for economic reasons. That left good old carbon. Diamond, with a particularly dense aggregate of atoms, would be an ideal moderator, but it is too expensive (even as synthetic diamonds). So graphite was the second choice. It is thermodynamically stable, chemically inert, very heat resistant with a good theoretical density of 2.26 g/cm^2 (although, in practice, graphite manufactured on a large scale has a density of only 1.7 to 1.8 g/cm^2). It needs to be purified as it is often mixed with metals such as boron. Typically, nuclear graphite is made from coke rather than petroleum to obtain a more isotropic material. (Anisotropy favours the apparition of cracks and reduces the service life of the moderators, which is still about forty years.) After an initial carbonization, the product is annealed, impregnated several times to increase its density and purify it, and then moulded. But more than developing the manufacturing process, the research was aimed at combating the damage that irradiation and high temperatures cause in the porous structure of graphite.[4]

These technical characteristics are not the only reason why graphite was chosen. Graphite blocks were initially used in the first atomic pile developed in Chicago in 1942. After the Second World War, France developed gas-graphite reactors using natural uranium – only the United States had enriched uranium – which released plutonium. Since plutonium is necessary for the manufacture of atomic weapons, it is therefore the alliance between civil and military nuclear power that makes graphite a nuclear material of strategic interest. Graphite is the key player in a dual technology, openly civil and fundamentally military. In France, graphite was useful for technological patriotism, enabling the country to earn a place in the concert of nations by asserting itself as a nuclear state.[5] In the 1950s and 1960s, France built eleven graphite reactors. Nuclear graphite saved France's honour, and contributed to its gallic pride, to the point that it claimed a place in the national heritage. The graphite-gas reactor of the EDF power plant in Chinon appears on postcards and wine labels in Touraine, as a symbol of the alliance of technological modernity with regional traditions.

Although nuclear graphite enabled France to assert its energic and military independence from other powers, it nevertheless lost some of its prestige in the face of economic pressures. After a long controversy between the French Atomic Energy Commission and EDF, the graphite-gas option was abandoned in 1969 in favour of the American light water system, the PWR (Pressurized Water Reactor). The end of the Cold War denuclearized graphite and put a stop to research on this allotrope of carbon, while the question of irradiated graphite waste remained unresolved.

Graphite thus had its moment of glory when it entered the ultra-high-tech, high-security and regulated world of nuclear power plants. However, its image was soon tarnished by protests against nuclear power, and its fate became even gloomier in the 1980s. Today, nuclear graphite is only of interest to the agencies responsible for storing the waste from decommissioned power plants. Moreover, the graphite used as a moderator in the RBMK[6] reactors, the second generation of Soviet Union reactors, proved to be unstable at low power. This defect was one of the factors in the Chernobyl disaster in 1986, when the reactor core melted. Will graphene be able to restore its image and reassert itself at the cutting edge of high-tech?

Graphene as an academic material

While nuclear graphite took centre stage in the techno-military arena, a more discreet – but no less strategic – existence opened up for graphite in the manufacture of intercalation compounds. Here graphite is depicted in the form of planes in which the carbon atoms, strongly bound by covalent bonds, are arranged in a hexagonal shape, while the individual planes are stacked with much weaker van der Waals bonds.[7] This results in a lamellar structure that allows reactants to intercalate by expanding the space between the planes. Owing to the chemical properties of graphite (both oxidizing and reducing), there is an exchange of electrons between the host and the intercalated guest. The foliated structure of graphite thus lends itself to the intercalation of all sorts of molecules, for example lithium, calcium, potassium or even metal alloys. These intercalation compounds are used in the manufacture of electrodes and batteries. In particular, graphite–lithium compounds form the negative electrode of the lithium-ion batteries used in mobile phones and laptops.[8]

Graphene was discovered in the earliest work on intercalation compounds. In 1962, German researchers described and measured 'very thin films of carbon' viewed through an electron microscope (Figure 9).[9]

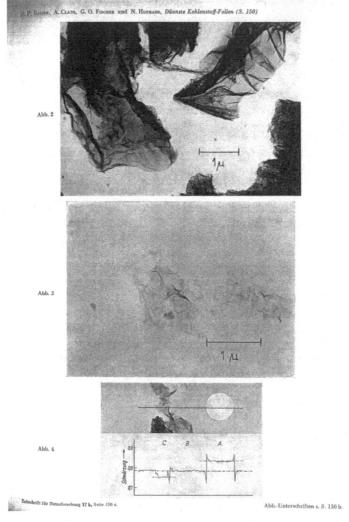

Figure 9. Graphene sheets isolated in 1962
© 1946–2014, Verlag der *Zeitschrift für Naturforschung.*

The author Hanns-Peter Boehm coined the word 'graphene' (composed of graphite and benzene) in 1980 and its definition entered the official IUPAC nomenclature in 1994.[10] But graphene remained above all a laboratory tool. It served as a theoretical model for understanding the stacking of sheets and electronic exchanges in intercalation. It also provided a textbook case to explain energy band theory, something every solid-state physicist needs to master.

The existence of graphene was considered purely theoretical, mainly for thermodynamic reasons. It was established as early as the 1930s that two-dimensional crystals cannot exist in a free state.[11] Once it reaches a certain size, the crystal minimizes its surface energy and small three-dimensional islands are formed. In short, the growth of a 2D structure is a physical impossibility. But this did not prevent researchers from growing graphene crystals in the laboratory, either by epitaxy on a metal substrate or in the form of rolls, cones or folded like origami, which minimizes their surface energy. They also managed to obtain graphene in the form of flakes in a liquid solvent, but in no case is it in a free state. There are no miracles on offer; graphite does not float upright, like Christ walking on water. Academic graphene is always deposited on a surface. It has a pure existence, in theory, and only impure incarnations in experiments where it is supported on a substratum.

A pure surface rich in promises

As with carbon nanotubes,[12] it is impossible to assign a date of birth to graphene. It was not discovered – or even isolated – in 2004 at the University of Manchester by Andre Geim and Konstantin Novoselov. Rather than its discovery, their 2010 Nobel Prize in Physics honoured their 'ground-breaking experiments regarding the two-dimensional material graphene'. So what experiments mark its coming into a new existence, graphene's renaissance?

At first sight, a harmless experiment, a 'Friday night experiment', a bit of pencil lead placed on a piece of tape . . . In his Nobel lecture, Geim jokingly recounted the 'random walk to graphene' of two Russian physicists, bored with the monotony of solid-state physics and in search of a more exciting science in 'Madchester'. With funding for a thesis, and a vague childhood dream (to move towards metallic electronics rather

than semiconductors), and noting that despite knowledge of intercalation compounds, not much was known about graphite lamellae, Geim
asked Novoselov to prepare graphite as thin as possible. Together with
a Ukrainian scanning tunnelling microscopy (STM) specialist from a
nearby laboratory and a roll of tape, these Russian emigrants 'peeled' the
graphite. It was Ukrainian microscopist Oleg Shklyarevskii who suggested
that they use tape to isolate the individual graphene film. Indeed, it is
customary for STMers to prepare the highly oriented pyrolytic graphite
(HOPG), often used as a support, by removing its surface layers with
adhesive tape, before throwing the piece of tape in the trash. But the tape
was not the sole thing contributing to the Nobel Prize.

The Manchester team had, on the one hand, built a small 'homemade' electronic device with graphene and demonstrated the great
'versatility' of its electrical properties, i.e. its ability to switch on demand
from almost zero conductivity to extreme conductivity (and inversely
for its resistance). The academic material thus became an interesting
technical device for the microelectronics industry.

Moreover, the Manchester team obtained a graphene sheet with
a surface area of a few square centimetres at room temperature.
Thermodynamically speaking, this was a metastable state of graphene,[13]
which compensated for thermal fluctuations by folding and undulating
gracefully in the third dimension. The Manchester team thus obtained
'freestanding' graphene, which can be transferred without alteration
from one substratum to another, retaining its stability under ambient
conditions and its electrostatic properties unaltered in the open air. It
could therefore claim to have understood the true nature of graphene:

> After all, we now know that isolated monolayers can be found in every pencil
> trace, if one searches carefully enough in an optical microscope. Graphene has
> literally been before our eyes and under our noses for many centuries but was
> never recognised for what it really is.[14]

What is graphene really? A 'pure surface', which no longer needs a
specific support to underpin it, which can be manipulated and used for
its own sake. This modest experiment opened the way to a new scientific
field: the science of two-dimensional materials. By detaching surfaces
from the volumes they cover, it allows for what Gilles Deleuze and Félix

Guattari called a deterritorialization.[15] It frees graphene from the hierar-chical relations in which it was entangled and allows it to reterritorialize itself in another context, that of nanotechnologies working at the limits of materiality.

At the limits of materiality

This level of electronic quality is completely counterintuitive. It contradicts the common wisdom that surface science requires ultra-high vacuum and, even then, thin films become progressively poorer in quality as their thickness decreases. Even with hindsight, such electronic quality is mystifying and, in fact, not fully understood so far.[16]

Why talk about 'quality' rather than 'properties'? The use of this unusual term by a physicist certainly reflects a desire to enhance the value of graphene and is often accompanied by superlatives: qualities that are exceptional, unique, extreme, etc. But beyond the hyperbolic effects, this term emphasizes the *qualitative difference* of graphene. A pure surface is not simply the edge of a volume, the gradual reduction of a dimension, it is *other*. It is not the surface of something but a surface in itself.

Furthermore, it would be instructive to compare the use of the word 'quality' by these scientists with the technical use of this term in classical metaphysics, which designates 'a mode of being' (*how* a thing is) as opposed to essence, which designates its nature (*what* a thing is). Seventeenth-century natural philosophers used to distinguish between primary qualities (e.g. Descartes's extension) and secondary qualities (e.g. smell, colour, texture, flavour), which do not belong to the object itself, but are a matter of sensation and are considered subjective. Primary qualities are the true properties of material things, those that belong to their essence – and it is generally in this sense that physi-cists speak of the 'intrinsic properties' of physical objects. So what is it about the electronic, magnetic, optical and chemical 'qualities' of graphene? Graphene is both a nanometric object that is invisible in one of its dimensions (its monoatomic thickness) and a macroscopic object that is sensitive, tangible and visible in the other of its dimen-sions (its surface). It can be manipulated and engineered at both these scales. As it is a pure surface, it would be difficult to distinguish between what is essential and what is accidental, what is deep and what is superficial, what is objective and what is subjective, sensorial or even social (the interests or values projected onto it). Graphene has an intrinsic technical value as a pure surface, a 'technicity' in the sense of Gilbert Simondon, which is close to aesthetic beauty.[17] The existence of a free-state 2D structure is both an ontological reality and a technical opportunity. It is as if the objective and the subjective, the physical and the social, no longer bifurcate into two distinct modes of existence but join together on the same surface.

The affordances of graphene seem to revive the dream of becoming free of matter, to the extent that this pure surface has all the makings of an ideal material that would combine all the functions at the highest level. It beats records for its ability to transport electrical charges and light, and for its lightness, which immediately attracts a host of high-tech industries. The intrinsic electron mobility of graphene in a free state is 200,000 $cm^2/V.s$, compared with only 1,400 $cm^2/V.s$ for silicon. Its planar structure makes its electrons so mobile that they behave like photons, massless particles. They can therefore be transported over long distances without loss, which is a dream for the microprocessor industry and high-frequency electronics (from a few tens or hundreds of megahertz to terahertz). Equally attractive for basic physics, it offers a pocket-sized device for doing quantum field electrodynamics. Optics, too, is interested in the ability of graphene to absorb light uniformly over a broad spectrum from infrared to ultraviolet.[18] It also offers possibilities for storing energy thanks to its conductivity and the surface area it covers; for manufacturing composites thanks to its lightness, which is greater than that of carbon fibres and allows aircraft to avoid lightning strikes because they act as Faraday cages; for designing flexible flat screens or conductive inks; for designing sensors by grafting chemical or biochemical probe species onto its surface; and for manufacturing cerebral or retinal implants (thanks to its conductivity and role as a membrane). In short, graphene fully meets the objectives of nanotechnology, which is frequently referred to as an 'enabling technology'. As an 'enabling material' *par excellence*, graphene suggests an indefinite increase and multiplication of our technical capacities. In this respect, it supplants carbon nanotubes because it is homogeneous. When one works on a batch of nanotubes, one always has a mixture of single and multiple walls, conductors and semiconductors with metallic impurities. With graphene, an almost absolute purity is achieved.

How can an almost immaterial material take on all the functions for which it is intended in microelectronics, optics, materials engineering, medicine, etc.? How can a pure surface carry the weight of all the investment Europe and other countries have put into it?

Unique and generic

What seems to distinguish graphene from all known materials is not so much its individual performances as the unique combination of the 'qualities' it offers. This combination fascinates and attracts specialists from all academic disciplines and industrial companies who – oxymoron intended – expect the unexpected, are foreseeing the unforeseeable: a technical and logical breakthrough with new applications that would change everything.

The resources (human, material, financial) now committed to speeding up commercial applications are impressive. Even if the Manchester team had demonstrated that graphene can exist in its free state, the quantity of ultra-high vacuum instruments, cleanrooms, microscopes and post-docs that populate the floors of the National Graphene Institute at the University of Manchester is a measure of the price to be paid to bring this material into existence and maintain it: heavy equipment, considerable investments and infinite care. Graphene is no longer produced with rolls of tape. Chemical vapour deposition (Figure 10) or the R2R (roll-to-roll) process is used, which makes it possible to produce single-layer sheets of about one metre in length.

Graphene has benefited from nano-scale measurement devices as well as from the know-how acquired in the chemical manipulation of carbon nanotubes. And since the objective of the roadmaps is to accelerate the transition from fundamental research to commercial applications, knowledge of this extreme carbon has progressed considerably in recent years, thanks to strong interdisciplinary initiatives.

Graphene no longer has the right to exist as an academic material; it cannot remain a material on the shelf. And yet this near-ideal material still has to prove itself. In particular, as the product differs greatly depending on its synthetic or fabrication process, setting standards for

Figure 10. Chemical vapour deposition of graphene © National Graphene
Institute, University of Manchester.

product regulation is a real challenge. Toxicity and enzymatic degradation tests give results that are difficult to generalize because they vary
from one product to another or from one protein to another. This is
because the graphene that goes into marketed products is never pure
graphene or 'in its free state'. For manufacturers, it makes little sense to
talk about a graphene monolayer, because the processes for extracting
graphene from graphite are statistical and there is always a proportion
of graphite in what is sold as graphene.[19] Depending on the application,
there is recourse to several modes of existence of graphene.

For most biomedical applications such as 2D delivery platforms,
what is used is not pure graphene but its hydrophilic form, graphene
oxide, which renders it possible to obtain graphene in suspension for
injections. In order to transport drugs to a target, it is advisable to use
several layers to obtain a certain permanence in the tissues before elimi-
nation by the urine. In biomedical applications, the focus is on flexibility
and sensitivity, whereas electronics requires an sp^2 distribution of pure
and very uniform graphene which is impossible to obtain by chemical

vapour deposition. But the race for purity is often only a prelude to the controlled introduction of defects or impurities to achieve the expected functions. The electronics industry made the effort to obtain silicon with a very high degree of purity, and then introduced dopants: a little phosphorus to produce electrons, a little boron to add holes. Similarly, attempts are being made to alter the perfection of the graphene structure in order to direct or trap electrons, or even to introduce magnetism, so as to extend the applications of graphene even further.[20] Battery electrodes, in particular, require defects, voids, cavities. The advantage is that, to create these defects perfectly, one does not always have to add dopants, as in silicon: the atoms of this 'pure surface' can be rearranged.

While graphene offers a unique combination of qualities that enable it to perform a variety of functions, it is neither perfect nor even optimal for its intended applications. It competes with other materials. And sometimes it is carbon versus carbon, graphene versus fullerenes and nanotubes. Graphene is thus a long way from dethroning the other modes of existence of carbon, in particular in the field of composites, where it is allied with carbon fibre. In a sense, the industrial development of graphene is a return to the art of mixing developed for plastics.

This means that the 'pure surface' mode of existence is largely a myth. But industrial development has led to a reverse movement from applications to basic research ('from brand to bench'). And the basic research clearly shows that although graphene is a unique material, it is also generic.

Graphene is now emerging as the head of a new family of two-dimensional structures. The same kind of performance can be expected in any material with strong in-plane bonding and weak inter-plane bonding. Graphene is thus becoming a textbook material that paves the way for other candidates for the heralded technological revolution. The research is focusing on boron nitride, molybdenum or tungsten disulphide and carbides, or metal sulphides or selenides. The aim is to design tailor-made materials, alternating layers of various 2D materials to meet specific functions. Hence the development, layer by layer, of heterostructures with interleaved dopants like in a custom-made lasagne. In these heterostructures, the graphene monolayers can change their crystal structure to realign with the boron nitride layers. This ability to self-rearrange is another example of how carbon responds to its neighbours. Developed with the help of 3D printers, these 'designer materials' are seriously fulfilling the dream of bottom-up design[21] that animated the technokitsch prophecies of molecular manufacturing.[22] Thus, after the nanotubes trend, graphene is becoming the iconic nano-technology material.

Does carbon hold the key to the 'technological revolution' and the power of nations? Fullerenes, nanotubes, graphite and graphene seem to be good candidates to foster all sorts of techno-utopias. It should not be forgotten that graphite and graphene were chosen as strategic materials not only by virtue of their intrinsic qualities, but undoubtedly also thanks to military, economic and technological conjectures. If these circumstances change, nanocarbons may lose their power to attract thousands of researchers and billions in investment. Thus graphene could well fade away to make room for competitors. It is an extreme material with exceptional, but not unique, qualities. Graphene has opened up a horizon of possibilities that could become commonplace if other materials were to exist as 'pure surfaces'. Graphene is perhaps a singularity that opens the way for tomorrow's technological commonplaces.

PART II

CARBON CIVILIZATION

9

Traces, stories and memories

We have seen how carbon was used by Mendeleev as a starting point for writing the periodic law of the elements.[1] But our shared history of entanglement with carbon goes far beyond its contributions to chemistry. Not only does carbon provide a molecular skeleton for all living organisms, but it is also the basis for the emergence of human cultures. Linked to fire by its very name, which means 'burnt', carbon gave our human ancestors the ability to inscribe traces. And it multiplies this capacity. Charcoal, pencil, carbon paper, diamond and radiocarbon are all ways of drawing, writing, engraving, dating and archiving. Carbon offers tools for sharing and fixing the present, transmitting it to future generations, reconstructing the past in the form of history and building chronologies. Carbon thus provides human societies with a power to control time.

Carbon as writer

In *The Periodic Table*, Primo Levi tells the story of a carbon atom.[2] He picks one at random, trapped in limestone ($CaCO_3$) for hundreds of millions of years, frozen in an eternal present where 'time does not exist', or barely just a few temperature oscillations. One day, this atom is suddenly cleaved off by a pickaxe and introduced into a lime kiln, precipitated 'into the world of things that change'. Now CO_2, it flies through the air, bound to two atoms of its former partner oxygen; for a few years it experiences a whirling existence; it is breathed in by a falcon, expelled, dissolved in the water of a torrent, evaporated again. One fine day, it passes into the bloodstream of an animal, and finds itself caught up in an 'organic adventure' where it changes identity and partners according to the vast and complicated molecular architectures it helps to form. Soon it is back in the open air, fixed in a leaf by a ray of sunlight; it then experiences the 'refined, minute, and quick-witted' chemistry of photosynthesis, 'whose scale is a millionth of a millimeter, whose rhythm is a millionth of a second'. It does tricks in the air, introduces itself into a cedar of Lebanon, experiences a benzene existence, is absorbed by a woodworm, passes into the earth, then slowly rises in the sap of a vine before knowing the mysteries of wine fermentation. Again, a new chemistry, where our carbon enters a glucose molecule, soon oxidized. Eventually it enters a human body. The blood flow carries it to a neuron, in a brain that is none other than that of the writer, whose hand holds a pen that traces black signs on paper to inscribe the final full stop of his book.

The chemist-writer[3] Primo Levi transformed the periodic table into a novel, the classification into a narrative.[4] Why is carbon the story on which he ends? Because carbon participates in writing, as if it were passing from the author's brain to the ink impregnating the fibrous texture of the paper. The writer narrating and the carbon being narrated confuse their roles. They merge in the very act of writing in its materiality.[5] 'The cell in question,' writes Levi-carbon, 'and within it the atom in question, is in charge of my writing, in a gigantic, miniscule game which nobody has yet described.'[6] And Levi-carbon marks a final full stop, which is also a point of intersection between material inscription and the inscription of meaning, a point where natural history flows into cultural history, a point that concentrates all the common history we share with carbon: that of *writing*.

Carbon as graphic designer

Since the dawn of humanity, charcoal and carbon black have been among the main pigments used to inscribe pictograms, outline animal figures and highlight rock reliefs on the walls of caves decorated by our distant ancestors more than thirty thousand years ago. Carbon thus provided humans with their first writing tools, their first graphic media.

Carbon black from smoke and soot is also the traditional pigment for most black inks, such as Indian ink. These inks were used long before calligraphy to mark bodies, as evidenced by the traces of carbon black tattoos dating from the Neolithic period found on Ötzi the Iceman, from five thousand two hundred years ago.[7]

Charcoal drawing is another of the well-known coal-based graphic arts. It lends itself to shading, modelling and texturing, and was particularly popular in the nineteenth century, but its artistic as well as utilitarian use has been documented since the Renaissance. Less well known is its traditional manufacturing process. A French eighteenth-century *Dictionnaire des arts et des sciences* tells us that:

> Painters & Engravers use *Charbon de Garais* to make their sketches. It is made by putting willow wood in a gun barrel and then heating it in a fire, converting it into Charcoal. The fruit of the Garais is square with four cores; when it is cooked in a barrel it is called *Fusin*.[8]

Garais, guéret or *bonnet carré* (because of the square shape of its fruit) is the French name for a shrub with pink berries, European spindle or *Euonymus. Fusin*, or in its modern spelling *fusain*, is the French name for a charcoal drawing tool. *Fusain* thus takes its name and shape (a regular cylindrical stick) from its preparation in gun barrels, where willow or spindle wood was charred – hence its older, other name of *Charbon de garais* (spindlewood charcoal).

The pencil, consisting of a graphite lead inserted into a hollow wooden stick, was probably invented during the Renaissance. But unlike coal black or spindlewood charcoal, carbon hides its name in pencils. Since the ore extracted from the Borrowdale pits in Cumberland in England from the 1560s onwards was much purer, denser and brighter than the coal extracted in Wallonia for heating and tempering iron in forges, it was associated less with coal than with lead or antimony and was given various names, such as *English antimony, kish, kellow, killow, wad, wadt* (meaning 'black' in some English dialects), *black-cowke* . . . but it was mainly the names associated with the metallic element lead that were retained, such as black lead, or plumbago (derived from *plumbum*, 'lead', with the suffix *-ago*: 'which acts like lead'). The English name of the writing part of the pencil is in fact 'lead', whereas in French it is called *mine*, or sometimes *mine de plomb*, a name that refers to the place where the ore is extracted.

Why lead when it was graphite? This isn't because the chemists of the time were shoddy scientists who mistook one substance for the other. In the 1690s, the naturalist Robert Plot analysed these mines and concluded that the substance in question was neither a metal like lead (because it was neither fusible nor ductile) nor a stone (because it was not hard enough), but rather an earth that was not soluble in water. This is why he called it *Ochra Nigra*, or black ochre.[9] Only the black colour mattered. In fact, it was for commercial reasons that the terms 'black lead' or 'plumbago' prevailed. This was to compete with the real lead *mines* or 'lead styluses' commonly used since antiquity, but which were quite expensive due to the scarcity of lead ore. In order to sell the pencil, it was presented as a mineral that *acted* like lead (plumbago) but was *blacker* than lead (*black* lead).

In his book *The Pencil: A History of Design and Circumstance*, Henry Petroski links the history of the pencil with that of engineering.[10] The pencil is one of the main tools of the classical engineer and it belongs to the engineering arts. Unlike ink, the pencil is self-effacing; it enables and narrates the creative process. It was conceived mainly to draw. Through successive repetitions and sketches, the drawing becomes a *design*.

Petroski emphasizes how the pencil lends itself, far more than the lead stylus, to constant engineering in terms of process and usability. This is primarily due to the properties of the material. To become a writing tool, plumbago requires invention. It can be mixed with other substances to obtain a more or less contrasting, dry or greasy, matt or shiny finish, which is only possible with lead at the cost of making metallurgical alloys. Lead must be melted and machined. It requires rigour.[11] Lead styluses are distinguished from each other primarily by their ornamental qualities: from the simple nail-like point to the stylus with a richly decorated head. The different ranges of pencils, on the other hand, are distinguished by their technical qualities. They are designed for a two-fold adaptation, to the user and to the paper, and invite improvements in both directions. Made from a single piece, the lead stylus remains fixed in time, not changing much and for a long time less expensive than a wooden pencil. Although it requires rigour, lead remains malleable and ductile. Its trace is thicker but also duller (grey to dark grey). The pencil is finer, more contrasted, blacker, prouder. But it can be thicker or lighter, if desired. The lead stylus sharpens like a blade, and the pencil becomes progressively smaller and needs to be gently sharpened as it is used.

In terms of drawing, lead lends itself to sharp, sure and precise contours; pencil lends itself to sketching, blurring and reworking of contours, but also to hatching and shading. The pencil can be adapted to different ways of drawing. But above all, its trace can be better erased than that of lead. Both could be faded with breadcrumbs soaked in milk, but to erase lead marks, the paper had to be scraped with a blade. Lead demanded rigour and invited style. The pencil encourages invention by inviting re-conception. Its first users were craftsmen – artisans, planners, cartographers and engineers – whereas for the lead stylus they were men of letters. Tracing with a stylus required confidently expressed thinking. By contrast, using a pencil was – and still is – to let oneself be guided by traces, to think in the immanence of traces.

If, for a long time, craftsmen and engineers were men of the pencil and scholars men of the stylus or lead, later, with the mass reproduction of books, the pencil took pride of place in the practices of scholars as a tool for annotation, i.e. for reading/writing. The first latex erasers were marketed in England in 1770 by Edward Nairne, a manufacturer of optical, barometric and electrical instruments, at the prohibitive price of three shillings per half cubic inch, since the rubber had to be imported from South America. They were democratized in the nineteenth century and confirmed the triumph of the pencil over lead. As for today's 'lead pencils', they are graphite pencils. But graphite remained very expensive and did not dethrone lead until the nineteenth century. Schoolchildren in the eighteenth century still ruled their notebooks using actual lead pencils.

Moving from a lead identity to a carbon identity, 'lead' production for pencils became international and it was no longer a monopoly of English mines. In 1779, the Swedish chemist Carl Wilhelm Scheele analysed the plumbago he used for writing and showed that it was identical to that of a very dense mined coal. Then, in 1789, the German mineralogist Abraham Gottlob Werner named it 'graphite' from the Greek verb *graphein*, meaning to write.

But just as the genie was emerging from its wooden sheath to reveal its coal identity to the world, the French Revolution broke out. At war with a coalition of European monarchies, the French nation was subjected to an economic blockade that deprived it, notably, of pencils. Given their importance for the – primarily military – arts of engineering, the issue was highly strategic, as if today's engineers no longer had computers. The Convention government consulted the Agence des Mines with a view to obtaining a substitute for English plumbago in the manufacture of pencils. Lazare Carnot, a mathematician and member of the Convention and of the Committee of Public Salvation [Comité de salut public], where he had the army portfolio, entrusted a popular scribbler-inventor by the name of Nicolas-Jacques Conté with the development of a new process for the manufacture of pencils. After a few days' work, Conté had the idea of reducing coal to a fine powder to remove impurities. He thus obtained graphite powder, which he mixed with clay and water to create 'lead'. These are fired in coal. The leads are then placed in a rectangular housing closed by a wooden slat. The whole thing is finally machined to obtain the desired shape (cylindrical or hexahedral). The lubricating power of graphite gives the 'lead' its greasy side, while the clay, a small proportion, gives it its dry side. After obtaining a patent in 1795, Conté built, in less than a year, the pencil factory that bears his name. Although criticized for being inferior to English pencils, the Conté pencil was adopted by artists and the process was soon adopted throughout Europe. Thanks to lower production costs, the graphite pencil decisively left lead styluses far behind.

Diamond as engraver and reader

With diamond, we discover another writing carbon, but in a diametrically opposed way to graphite. Graphite leaves a trace by crumbling. As it is inscribing, it gives a little of itself. Diamond, on the other hand, gives nothing away. Nothing resists it; it scratches everything and is not scratched by anything. Precise, clean, effective; that is the diamond signature.

Given this, diamond volunteers itself as the *universal engraver*. There are countless tools and machine tools that use diamond to engrave, incise, indent or cut: chisels; milling machines; diamond turners and cutters adapted to hard and brittle materials; diamond scalpels for micro-surgery (especially of the eye); or diamond-tipped penetrators used for hardness measurements by the Vickers process (which measures hardness based on the size of the indentation left by the diamond in a material). Finally, in jewellery, the only material capable of machining or polishing diamond is . . . another diamond.[12]

So diamond certainly lives up to its name – it is adamant (Gk: ἀδάμας), the untameable, the inflexible. But is it really the hardest material ever known, going by its reputation? Hard it certainly seems, since traditionally the hardness of a material is defined by the resistance of its surface to the penetration of a diamond punch. But when it comes to resistance to stresses – tensile strength, contortion, shearing, bending – diamond is beaten by nanotubes, which in turn are beaten by flexible graphene, even though it is nothing more than a layer of the brittle and breakable graphite.[13]

Diamond is classified by materials science as an 'intrinsic superhard'. This group includes materials that are naturally hard in the solid state and have a Vickers hardness of more than forty gigapascals, such as diamond, carbon nitrides, cubic boron nitrides and rhenium nitrides, some of which outperform natural diamonds in terms of hardness. Natural diamonds, which are so valuable in the luxury sector, actually vary greatly in hardness. The more defects they contain in their structure, the less hard they are. In this respect, synthetic diamonds produced at high pressure and high temperature outperform natural diamonds and all other intrinsic superhard materials with up to a hundred and fifty gigapascals of Vickers hardness. But diamond, whether natural or synthetic, is still combustible.[14] So it can't be used to cut ferrous materials at high temperatures, including steel. Current research into superhard materials is therefore focused on developing compounds that are more thermally and chemically stable than pure diamond.

But there are also '*extrinsic* superhard materials', whose extreme hardness is due more to their nanometric shape than to their chemical composition. Some of these materials outperform diamond, and they are . . . other diamonds. In particular, a recently synthesized material known by the name of 'aggregated diamond nanorods' appears to be the hardest of the superhard materials to date.[15]

In short, diamond may have been dethroned, but this does not diminish its hardness. On the contrary, the hardness has become contagious. How are these astonishing superhard materials – including nanodiamonds that are harder than diamond – synthesized? Often in diamond presses. The aggregated diamond nanorods are obtained by compressing fullerenes between two diamonds pressed together at titanic pressures of several hundred gigapascals.[16] With diamond, solid-state physics fully deserves its name of 'condensed matter' physics. Thus diamond continues to write the history of matter and materials. This superhard material, the engraver of hearts and materials, also has many other ways of being precious.

A more familiar device illustrates the engraving skills of diamond in an original way. Phonograph and phonogram refer respectively to the writing (*graphein*) and recording (*gramma*) of the voice (*phonè*). In this apparatus, diamond plays a dual role: on the one hand, as a tool for engraving the sound in a matrix of metal, wax and then shellac (phono-graphy); on the other, as a tool for reading the recorded sound (gramo-phone). The two separated from the 1920s onwards, the phonograph being reserved for the production of sound objects, and the gramophone for listening (it could no longer be used to record). For a time, the latter abandoned diamonds, leaving them for engraving, in favour of cheap needles, which had to be changed after a few plays. They were replaced by sapphire styluses, which were more durable, and then by diamond styluses in the 1960s, which accompanied the rise of pop music.

With diamond, the quality of the playback sound is close to that of its writing. Placed at the end of the stylus, it transmits the deviations in the groove to a piezoelectric transducer which transforms these vibrations into an electrical signal; an amplifier then recovers the extremely low mechanical power transmitted by the faithful diamond. Without it, there is no 'high fidelity', no 'hi-fi system'.[17]

In the series of carbons that write sound, we must pay tribute to the carbon microphone. This is a mechanical-electrical transducer invented in the 1870s. Carbon granules are inserted between two metal plates in a capsule closed by a flexible membrane; an electrical voltage is induced between the two plates; the sound wave causes the membrane to vibrate, which compresses the carbon granules and causes a variation in resistance modulating an electrical signal. At the 'end of the wire' (or wave), another carbon transducer operates in an electromechanical amplifier mode, demodulating the electrical signals and reproducing the sound. In electrical circuit diagrams, it has this symbol:

Affordable, and with a good frequency response range, the carbon microphone/amplifier was for decades a faithful device for capturing, transmitting and amplifying the human voice, despite a slight hiss, the sound of crumpled coal whispering in our electric ears. It was used in most telephones (but also crystal sets, vacuum tube radios, and microphones at the BBC) until the 1960s, when it was replaced by germanium and silicon transistors.

Radiocarbon dating

Carbon-14 was isolated in 1940 in the cyclotron of the Radiation Laboratory at Berkeley for use in radiation therapy. Shortly after the detonation of two atomic bombs on Japan in 1945, Willard F. Libby, working at the Institute of Nuclear Physics at the University of Chicago, freed himself from military service to work on the development of a dating method using this radioisotope of carbon. He submitted a paper to *Physical Review* in 1946 on the detection of minute traces of radioactivity. For two years he trained to isolate a few millicuries of carbon-14. In 1947, he finally shared with his collaborator James Arnold the clever idea suggested by three modes of existence of carbon: ^{12}C, ^{13}C, ^{14}C. His method took advantage of the respective proportions of these three isotopes to determine the date of death of samples of organic matter. At the time, it was based on very fine measurements of very weak radioactivity signals with the aid of a counter and a robust hypothesis. The basic assumption, which has not been disproved, is that during their lifetime organisms constantly exchange carbon with their environment and statistically contain the same proportion of carbon-14 (^{14}C) as the rest of the biosphere. When the organism dies, it no longer absorbs carbon and the ^{14}C it contains decays following an exponential decay law. As the half-life of ^{14}C is 5,568 years,[18] the amount of ^{14}C in the sample halves every 5,568 years. It is therefore possible to deduce from the ratio $^{14}C/C_{total}$ in a sample the approximate date of death of that organism, whether plant or animal.

While he was on vacation, Arnold told his father about the idea, who repeated it to some friends. As a result, in May 1947, Libby saw samples arrive on his desk from the director of the Metropolitan Museum in New York.[19] Surprised, Libby decided to take a more open approach to collaboration. He contacted a senior colleague, Harold Urey, a Nobel Prize laureate in physics and a specialist in isotope separation, who himself formed an alliance with the physicist Harrison Brown and Robert Redfield, an anthropologist from the University of Chicago. This trio actively lobbied to interest archaeologists in the potential of this new dating method. Libby presented his technique to archaeologists and this first presentation paved the way for a series of regular seminars from 1948 onwards. A Radioactive Carbon-14 Committee under the American Anthropological Association was created to facilitate exchanges between the anthropology and archaeology communities and the nuclear physics community. Archaeologists provided samples already dated by conventional methods,[20] which allowed Libby and his collaborators to test their own method. The first test carried out in 1949 was on a piece of wood from the Djoser burial complex at Saqqara in Egypt, which was dated by ^{14}C to 4,650 years ago, plus or minus seventy-five years BP.[21] The dating process aroused enthusiasm among archaeologists, who believed they had acquired an almost magical procedure. The second test on a supposedly more recent sample gave a nil result and Libby ignored it for almost twenty years (without suspecting that he might have been handed a fake to test him). So the method was nothing like a magical tool for any archaeologist to use. It raised many difficulties because one has to detect minute traces, especially on very old samples, and the counters, which are often out of order, must be constantly recalibrated. But these difficulties, instead of discouraging the archaeologists, seem to have encouraged them to collaborate. Far from fearing the intrusion of conquering physicists into their own backyard, anthropologists and archaeologists have instead seen carbon-14 as a means of enhancing the image of their discipline. Carbon thus succeeded in bringing about dialogue and cooperation between two communities – the hard sciences and the humanities – at the very time when the physicist C. P. Snow deplored the total absence of dialogue and sympathy between the two cultures, scientific and humanistic, in his famous Cambridge lecture.[22]

Of course, the carbon dating method has its limitations. It can only be applied to samples of organic matter that must be kept contained.[23] It is not reliable for samples older than thirty-five thousand years. Beyond that, carbon must be replaced by potassium/argon or rubidium/strontium. But the method developed by Libby has been greatly improved thanks to mass spectrometry, which allows work on smaller samples in less time. Its more serious limitation lies in the many corrections made necessary by the difference between the ^{14}C content of the atmosphere today and in 1950. Just as the nuclear tests of the 1950s had enriched the atmosphere with radiocarbon, so carbon dioxide from fossil fuels has depleted it.

The carbon archive

In addition to inscription and dating *tools*, carbon also offers inscription and archival *media*.

Carbon paper is a sheet coated with a fatty substance and carbon black. It is used between two sheets of plain paper to make one or more copies of an original document. Invented simultaneously and independently at the beginning of the nineteenth century by the Englishman Ralph Wedgwood (for the stylus) and the Italian Pellegrino Turri (for a dedicated typewriter), carbon paper was not initially intended for copying, but for *invisible writing*. It allowed writing without ink and therefore without leaving a visible trace. Legend has it that Turri fell in love with a blind countess, Carolina Fantoni da Fivizzano, and built a machine for her to correspond in private. In 1808, the Countess wrote that she was desperate because she would soon find herself without black paper. Turri was her only supplier.[24] It was from the 1870s onwards that carbon copying became standard practice for duplicating documents, contracts and administrative receipts of all kinds.

But carbon paper has two disadvantages. It allows only a limited number of copies and can only reproduce outgoing correspondence. If a company needed to copy incoming documents, they had to copy them by hand onto the carbon paper. As businesses continued to grow, other duplication processes were sought that could produce more copies by hand, such as cyclostyle, stencil, mimeograph and various alcohol copiers. Carbon had not been dethroned, however. These copiers, familiar to schoolchildren a few decades ago, included carbon paper, the ink of which was transferred to paraffin paper, which was then soaked in alcohol as it passed through the rotary press. Its printing, with the characteristic smell of alcohol and petroleum, was purplish because these machines used aniline purple as a dye. The stencil, on the other hand, duplicated without carbon paper but did not solve the problem of copying incoming documents. Only the Xerox photocopier, marketed from 1950 onwards, slowly but inexorably began to reduce the demand for carbon paper. Carbon paper has almost disappeared from today's computerized offices,[25] but not without leaving its *mark*. The abbreviation 'cc' stands for 'carbon copy' and refers to the duplication of electronic messages in as many copies as there are recipients. 'Bcc' refers to the same function but for recipients who are invisible to each other (*'blind* carbon copy'), in a way fulfilling the dream of the inventors, Wedgwood and Turri, on a larger, worldwide, scale.

Finally, even though computers have chosen silicon for computer memories, carbon could well compete with it. There is talk (or a dream) of graphene memories,[26] or even of a new electronic, analogue form of writing, based on frequency coding.[27] In a way, this would be the graphite pencil's revenge on digitization.

But carbon has already provided a *code*: DNA. It is built around a carbon skeleton like all the other basic molecules of life (proteins, carbohydrates, lipids, etc.). Living things are written, and writing, in the form of sequences of letters, all composed with the help of carbon. Now, DNA is indeed a *sequence of grams* (ATGC), alongside the four others of RNA and the twenty-two amino acids. It even makes *anagrams* (mutations, polymorphisms, reassortments) and *engrams* (it memorizes traces and updates them according to encounters with phages and viruses, pathogens and partners). And current research into the storage of information in synthetic DNA molecules suggests that carbon will take over from silicon in the construction of memories capable of storing 215 million gigabits in one gram of DNA.[28]

At the end of this somewhat dizzying journey through literature, writing, music and archaeology, it is clear that the multiple identities of carbon have much more in common than chemical allotropy. Allotropes, isotopes and compounds of carbon each develop in their own way a relationship to time and memory. And overall they offer a fix on time in two ways: the *graph* and the *gram*. Put simply, the former is on the side of writing, the latter on that of reading. As graphs, by inscribing material traces, they provide tools for writing, engraving, drawing and marking. As grams, they record and provide memory platforms. From coal to pencil, carbon is clearly on the side of the graph. Graphite is a graphic artist but its trace is ephemeral: it must be fixed with lacquer or preserved in caves that have remained empty and lifeless for thousands of years before coming down to us. More often than not, it fades away. Diamonds, on the other hand, are engravers, recorders and readers of grams, and thanks to their hardness, they are inscribed over time. But to inscribe things in time, we also have the grams of DNA, RNA, in short *soft carbons*. Carbon-14 marks time like a slow hourglass that loses its isotopes, breaking off one by one. Carbon thus links material culture and cultural history. It provides the essential infrastructure for the various human cultures and creates a bridge between these cultures. What follows will show that history continues to multiply our dependencies on carbon.

10

The resilient rise of fossil fuels

It is no secret that the issue of fossil-based energy is at the heart of the global economy and politics. It drives international relations and, despite decades of warnings, global fossil fuel consumption continues to grow. The grip of fossil fuels on us is so strong that we see it as fate or damnation, as the price to pay for progress, for comfort . . .

How did carbon compounds manage to take over so much power, to the point of pulling the strings of social and political history, and putting the future of the planet at stake?

Memories of life on earth

Fossil fuels are like a record of the past, the archives of life, its material memory compressed in the soil. The slow sedimentation of layers of biomass transforms it into necromass, and under the effect of heat and pressure, these layers write the memories of life on earth.

Petroleum – which will be the subject of the next chapter – is formed from the organic matter of algae, plankton and small animals that has settled to the bottom of oceans and lakes. The oldest are around five hundred to six hundred million years old, the period of mass sedimentation. They originate from the formation of kerogen, the threads of organic matter trapped in a mineral matrix, the parent rock, which were first affected by bacteria and then by rock. Over geological time, the kerogen undergoes the action of pressure and temperature, which increases as it sinks. It releases oils (crude oil) and gases (CO_2 or hydrocarbon gases). Oil can in turn be caught up in the movements of plate tectonics; its deposits can sink deep underground and undergo deep pyrolysis or partially rise; trapped below the surface, they form pockets of gas or turn into tar sands as they are broken down by bacteria. The diversity of oils is as great as the living things from which they are derived. No two oil fields provide exactly the same oil.

The oldest coals date from the so-called 'Carboniferous period', which began three hundred and sixty million years ago when terrestrial plants colonized the earth's surface in abundance. A portion of them accumulated in anoxic areas such as large swamps. Depending on the geophysical conditions, the turning of the soil transforms them into peat, coal and then either coal gas or lignite and then anthracite, which is almost pure carbon, stripped of most of the hydrogen coming from organic matter.

Current research at the crossroads of geology and physics is shedding light on the fossilization mechanisms that transform organic matter into coal, coke or graphite. Jean-Noël Rouzaud, a leading carbon specialist at the CNRS [Centre National de la Recherche Scientifique], explains that we can simulate the process for fifth-grade children. If we take, for example, pine wood and carbonize it, we first release water, carbon dioxide and then hydrocarbon compounds that form soot. The product obtained is a charcoal, richer in carbon than the material we started with. In a more professional way, the processes of coalification or graphitization that take place in the bowels of the earth are also simulated in the laboratory, and the structure of the products obtained is analysed using transmission electron microscopes and Raman microspectrometers. The more or less ordered structure of the charcoal depends on the nature of its precursor (more or less rich in oxygen and hydrogen) and its state (solid, gas). It also depends on the temperature and pressure. Between 500°C and 1,000°C, it goes through a plastic stage, coke, which is used in the steel industry and costs twice as much as coal. It owes its reducing properties to a certain balance in its structure between order and disorder. If we continue maturing, at 2,800°C we obtain graphite, but if we increase the pressure, we can obtain graphite sheets at around 1,000°C and, at a very high pressure of ten gigapascals, diamond. Thus, a wide range of carbons that nature has formed over millions of years are deployed within just a few hours. The structure of each one bears the signature of its formation at the same time as it signs a kind of utility contract with humanity.

Fossil hydrocarbons are so full of energy thanks to the strong carbon bonds inherited from living things and compressed by rocks and soils. This material memory of life on earth is formed over hundreds of millions of years. It is renewed in tens of millions of years. Yet today we are expending several of those millennia each year to exploit the energy of carbon bonds and to fuel industry, releasing as much CO_2 into the atmosphere. In this way, we are consuming and burning the memory of past life; we are burning 'buried sunshine', in the words of ecologist Jeffrey Dukes.[1] The fate of fossil fuels illustrates the common history humans share with carbon: as Marx and Engels famously put it in their *Communist Manifesto* (1848), *all that is solid melts into air.*

A carbon liberation front?

The story begins in the second half of the eighteenth century with the so-called 'industrial revolution'.

It would be difficult to imagine this taking place using huge amounts of biofuels like wood. Slaves instead of machines? Massive exploitation of the Amazon Rainforest with the Portuguese leading the field? Without pushing this little exercise in counterfactual history any further, we can already guess at everything that depends on the choice of fossilized carbon. Carbon from living things would not have been enough to fuel the manufacture of iron and steel, for instance.

When it comes to energy, any freedom, any gain in autonomy, implies a new dependency. Bruno Latour, in one of his last writings, speaks of a 'freedom of dependence'.[2] While carbon flows freely – i.e. by virtue of natural necessities – from one kingdom to another, our energy choices weave together economic, social and environmental regimes in ways that are difficult to untie. Fossil fuels are hybrids of nature and society. Neither coal, gas nor oil spontaneously holds the power to drive industry and create economic growth. They do have the potential to provide power for aircraft and automobiles, but converting them into fuel requires mines, pumps, refineries, machinery – a whole socio-techno-economic network of extraction, transportation, transformation, manufacturing, production, distribution, sales and monitoring. Among the many candidates for the function of the main driving force, some manage to achieve this thanks to a subtle mix of intrinsic properties, dispositions and affordances. Fossil fuels have been established for two centuries, have persisted and still persist because these networks of production, communication and alliances have been established and are still maintained – financially, logistically, strategically … It is a world woven of bonds of solidarity, but also of exclusion, where potential rivals were sidelined and protests were silenced. In short, could the last two centuries be the work of a carbon liberation front?

This is the provocative reading of the Anthropocene by McKenzie Wark, author of the hacker manifesto *Molecular Red*.[3] Riffing on Marx and Engels' 'all that is solid melts into air', Wark imagines that the massive recourse to fossil fuels is the work of a political liberation movement; not of the proletariat, nor of women, but of carbon molecules. A kind of secret conspiracy pushing the levers of world politics would have undertaken to release all the carbon on the planet in the form of gas into the atmosphere. By conjuring up the figures of a few forgotten Soviet pioneers and imagining their encounter with the Californian high-tech avant-garde, Wark offers a fanciful and at the same time terribly powerful reading of the real relationship that has developed between humans and carbon notwithstanding the rhetoric about carbon neutrality. She mobilizes the critical power of the counterfactual narrative to draw attention to the entanglement of technology, politics and economic interests that underpins the fossil fuel empire.[4]

Multiple coals

Hybrid in terms of nature, economy and society, fossil carbons share this condition with other materials. Just as concrete, plastics and composites came into the world to do a job as substitutes for another material, fossil carbons exist in the mode of potential substitute for another fuel. Some historians explain the transition from an agrarian system, dependent on solar energy, to an industrial system, dependent on fossil fuels, through a 'wood crisis', resulting from massive deforestation, itself attributed to limited agricultural resources, demographic pressure in Europe and/or falling temperatures.[5] In fact, mineral coal was first referred to as *Sylva subterranea* (subterranean forest) because it was seen as a substitute for forest wood.[6] The empire of coal became so naturalized that, at the beginning of the twentieth century, people spoke of 'white coal' to refer to hydraulic energy, forgetting that water mills had powered the textile industry before steam engines.

When one source of energy takes the place of another prevalent source, it gains the power of a lieutenant maintaining order and the smooth running of the troops. But as substitute, this entity is reduced to its function as a source of energy. It loses its singularity and its links with its place of origin, with a milieu. And it is only insofar as it is considered as a substitute that it can mark an era or the end of an era. Indeed, just as materials mark human prehistory – the Stone Age, the Bronze Age, the Copper Age – hydrocarbons periodize history. Coal is associated with the industrial revolution and the end of the 'oil epoch' is announced. However, this kind of periodization and the repeated invocation of 'revolutions' (the first, the second . . . the umpteenth industrial revolution) mask more complicated relationships between the economy and materiality.

Just as iron has not completely displaced wood for building, so coal has not completely taken over from charcoal. It did not immediately dominate industrial activities because of the shortage of wood. Although coke – obtained by distilling coal in the open air – was proposed as a substitute for charcoal in blast furnaces as early as the beginning of the eighteenth century, the steel industry did not immediately adopt it, despite warnings about deforestation. As Gérard Borvon rightly points out, charcoal was much more valuable than coke, which required heavy processing of mineral coal, including a series of batteries for distillation.[7] In contrast, the manufacture of charcoal is much easier. The carboniz-ation process involves stacking wood in overlapping beds to form a pile called a mound or kiln (Figure 11). To prevent or limit air access, the pile is covered with an envelope of dry leaves and moss, topped by a blanket of soil mixed with dust from previously charred kilns. The fire is lit from above, by means of a chimney in the centre, or sometimes through channels dug in the mound near the ground. The charcoal maker controls the charring by regulating the ventilation. However, not only does steel-making require fuel for the blast furnaces, but the iron must also be 'cemented', i.e. carbon must be allowed to penetrate to the core. For this operation, the naturalist René-Antoine Ferchault de Réaumur recommended, in the 1730s, a mixture of charcoal dust with soot, ash and sea salt.[8] This use of charcoal in steel-making is still in force today in northern Brazil, where it is contributing to the devastation of the Amazon Rainforest.

On the other hand, materials that we treat as mere substitutes for a given function – for example, as fuel – have many other modes of existence. Thus, in the eyes of naturalists, coal is a trace, a memory that

Figure 11. Preparation of charcoal by carbonization, Duhamel du Monceau,
'De l'exploitation des bois', 1764.

bears the imprint of its genesis. Long before they had access to the fine
structure of organic molecules, eighteenth-century scientists were able to
distinguish varieties of coal, so numerous that they defied their efforts
at classification. There is no standard coal, detached from a place, or a
history.[9] Every coal has its earth, is attached to a place, and has qualities
deriving from its original *terroir*. Like wines, coals are products of specific
territories. They are defined by their place of origin, which determines
their use. For fertilizers, the academician Henri Louis Duhamel du
Monceau particularly recommended the use of a 'blackish-coloured
earth', a kind of peat, which is found mainly in the marshes of Holland
or along the canal from Lille to Douai. The texture, smell, colour and
combustibility of the coal are indicators of its composition, which varies
greatly from place to place as it depends on the time and conditions of its
maturation. For each coal has its own history, and this history determines
its structure as well as its properties and functions.

The identification of fossil coal as a 'fuel' has nothing natural or necessary about it. It has come out of countless ventures into mining, trading, court cases and legislation. Its function as a driving force for industrialization was far from fixed at the beginning of the nineteenth century.[10] Mineral coal was used throughout the eighteenth century in the agricultural sector. In his *Éléments d'agriculture* (1762), Duhamel du Monceau presented 'fossil earth' as a fertilizer.[11] To prepare the fertilizer, water is poured over it and it is kneaded with the feet until it forms a dough from which 'cakes of 7 to 8 inches in diameter' are made. They are left to dry, while taking care to keep a little moisture, which is necessary for the soil to catch fire. Then a pyramid is made from these cakes with a fire pit filled with straw at the base; the fire is set and left to burn for two or three days. Once the ashes have cooled down, sixty to eighty pounds of them are spread per acre, with hands protected because these ashes are very caustic, and in April or May the earth will bloom again. Duhamel du Monceau specifies that one can also use the ashes of coal burnt in glassworks and breweries, even if it is not as good as peat from Holland. The important thing is that fossil coal is used to restore the balance of nature. In this way, it contributes to the *oeconomy*, in the sense of resource-saving household management (*oikos*). It is considered above all as fossil flora, and as a trace and witness of the earth's history, it is destined to return to the earth to fertilize it.[12]

The identity of wood charcoal is equally open. Let us not think that it was always the fuel used for barbecues or other cooking appliances. It is rather mineral charcoal that was reserved for domestic use for heating houses or cooking, as far back as the Romans. Furthermore, by perfecting the carbonization process, wood charcoal is also used to produce activated charcoal. After heating to 1,000°C to obtain a microporous charcoal structure, activation consists of 'opening the porosity', i.e. enlarging the pores with steam to obtain the desired pore size. This material, which can also be obtained from olive cores or peanut shells, has such interesting adsorption properties that it has not yet been superseded in various applications. Widely used during the First World War for the manufacture of gas masks, it is now used to capture toxic molecules in fumes. It forms the filter in factory or kitchen hoods. It is also marketed for water purification because the micropores serve as niches for microbes that clean up the effluents. The industrial manufacture of activated charcoals from willow, birch, lime, pine, etc., was established from the eighteenth century, but their medicinal and water purifying uses have been known for a much longer time. The Phoenicians, for instance, carbonized the water barrels of their ships to keep drinking water pure during long sea voyages.

Prometheus unbound

This means that coal is not just the driving force of the industrial revolution; coals have plenty of other affordances. The fact remains, however, that through its association with steam and the division of labour, coal has inspired the mythology of industrial progress, regularly celebrated in the Universal Exhibitions with galleries of ever more numerous, more powerful and more ravenous machines. The coal of the industrial revolution established a predatory and destructive economy that Lewis Mumford, with his critical eye as a historian of industrial societies, called 'carboniferous capitalism'.[13] Finally, it is coal that directs the great epic of the working-class world with its social movements and class struggle.

Coal's beginnings in industry were spectacular, if modest. In the early nineteenth century, William Murdoch developed a process for lighting with coal gas, a specific variety of mineral coal. He first illuminated a factory in Cornwall in 1807 and then perfected the process which was adopted for urban lighting in 1815. However, lighting public spaces is not the same as regulating the functioning of Western societies. How did we get to that point?

Firstly, industrialization is based on coal, but not on coal alone. The transition from an agrarian system that consumed little energy in the form of carbohydrates to an industrial system that consumed a great deal of fossil hydrocarbons was mainly due to a synergy of multiple technical innovations that mutually reinforced each other. The piston of Thomas Newcomen's machine was first used to pump water from mines. As the historian Andreas Malm points out, 'Only by coupling the combustion of coal to the rotation of a wheel could fossil fuels be made to fire the general process of growth: increased production – and transportation – of all kinds of commodities.'[14]

The patent filed by James Watt in 1784 to produce a continuous circular motion did not immediately pave the way for steam and coal engines. In England, the cradle of industrialization, steam engines did not replace water mills in the cotton factories near Manchester until the late 1830s. The steam engine became widespread especially when labour regulations significantly reduced the profits of cotton mill owners.[15]

With the energy-hungry steam engine, coal became a necessary part of the economy. Prometheus was unbound, according to David Landes's famous title.[16] 'Out of this coal and iron complex, new civilization developed,'[17] writes Mumford. As the economy was no longer exclusively dependent on agriculture, nor on the water of rivers to turn mills, the ruling classes felt that they were emancipated from the constraints of nature. An industrialist could store coal and no longer worry about seasonal variations. The concentration of fossil energy, accumulated in a few underground seams over millions of years, initiated a new regime of economic temporality:

> In the economy of the earth, the large-scale opening of coal seams meant that industry was beginning to live for the first time on an accumulation of potential energy, derived from the ferns of the Carboniferous period, instead of upon current income.[18]

This fabulous 'fossil capital'[19] may have led to the belief that indefinite growth was possible. The historian Edward Wrigley goes so far as to claim that coal would have enabled an escape from the vicious circle of growth that increased pressure on the environment.[20] For Mumford, on the other hand, 'mining is a robber industry: the mine owner . . . is constantly consuming his capital, and as the surface measures are depleted the cost per unit of extracting minerals and ores becomes greater. The mine is the worst possible local base for a permanent civiliz-ation.'[21] Coal mining promoted at the same time both the concentration of capital and the formation of the proletariat. Heavy investments were required to open and operate a mine. Investors expected profits in return and therefore sought to intensify exploitation and labour productivity. This exploitation fever was the basis for the sustained growth in the production of material goods, urbanization, railways – in short, the process of modernization. Mumford sees this as the origin of a social evil that persisted after the exhaustion of resources:

Mankind behaved like a drunken heir on a spree. ... The psychological results of carboniferous capitalism – the lowered morale, the expectation of getting something for nothing, the disregard for a balanced mode of production and consumption, the habituation to wreckage and debris as part of the normal human environment – all these results were plainly mischievous.[22]

On the other hand, coal also required a large concentration of workers, because despite the use of steam in the mine (for pumps and hoists), the miners attacked the coal seam with picks and shovels and horse-drawn wagons carried their produce. Novels like Émile Zola's *Germinal* (1885) or Richard Llewellyn's *How Green was My Valley* (1939) have fixed strong images in our memory, of dark villages around a mine attracting immigrants and gradually displacing the peasant world; the hard life of the miners' families with accidents, drinking, wages which the company cuts in the name of economic crises and strikes, demands, revolts. However, as Timothy Mitchell shows, coal offered the proletariat favourable conditions for mobilization and social victories.[23] Because coal was heavy and expensive to transport, industrial production was set up near the extraction sites, which concentrated the working-class population. Moreover, at the bottom of the mine, out of sight and in an extreme environment, the miners could develop solidarity that surface work did not offer. The intensive exploitation of coal was matched by laws limiting daily working hours and child labour. In short, coal made social achievements possible.

Scarcity foretold

Through its alliances with steam, capitalists and the proletariat, coal has profoundly marked Western societies. These have become increasingly distant from nature, with urban space detached from the countryside and time abstracted from seasonal cycles, regulated by the rhythms of production, rest and consumption. This industrial regime has such a self-reinforcing dynamic that it has been lulled into the mirage of limitless abundance and indefinite growth. In 1850, Britain alone consumed three times as much coal as Belgium, France, Germany, Austria-Hungary and the United States combined.[24] And Edward Hull's 1861 geological survey of its resources concluded very optimistically that Britain had eighty thousand million tons of mineable coal, which, it estimated, would sustain the 1850 rate of consumption for 1,100 years. They were riding high on optimism.

In 1865, the country was shaken when William Stanley Jevons published *The Coal Question*.[25] At the end of his own investigation, juggling geology with economics, Jevons announced a coal shortage. Whereas an economy based on the exploitation of the earth could be stationary, he explained, an economy based on a fossil resource was doomed in the long run. In his view, increasing the efficiency of machines through technical progress could be a way out. But saving energy by increasing efficiency paradoxically leads to an increase in demand and consumption, as the price of energy falls (what is now known as the rebound effect).

The book caused panic and controversy in Britain and throughout the industrial world. Prime Minister William Ewart Gladstone set up a Coal Commission and geologists were asked to make an inventory of accessible coal, as well as of hidden reserves,[26] not only in Britain but also in Japan, Argentina, Belgium and the United States. This major international geological inventory campaign resulted in a 1913 report that overestimated the world's reserves (except in the case of Russia).

The perennial techno-optimism of Jules Verne-type experts and engineers could not, however, hide the profound yet subtle change in the vision of the earth that the 'coal question' produced. The earth was no longer the well-managed common household, as in eighteenth-century *oeconomy*. It was no longer a habitat to be cultivated and maintained but a stock of resources to be evaluated, safeguarded and exploited.

A hoped-for turnaround

The question of coal shortage is even discussed in Jules Verne's *The Mysterious Island*, where the reporter Gideon Spilett expresses his concern that the steady increase in consumption leading to increased exploitation will sooner or later lead to the exhaustion of coal. To which the engineer Cyril Harding retorts: once the coal found in America or Australia is exhausted, water will be burnt. Water, broken down by electricity, will be the fuel of the future:

There is, therefore, nothing to fear. As long as the earth is inhabited it will supply the wants of its inhabitants, and there will be no want of either light or heat as long as the production of vegetable, mineral or animal kingdoms do not fail us. I believe, that when the deposits of coal are exhausted, we shall heat and warm ourselves with water. Water will be the coal of the future.[27]

With his unbridled techno-optimism, Verne's engineer points in a direction that could have been taken at the beginning of the twentieth century. Indeed, the 'hydrogen engine' was on track more than two centuries ago. Since the decomposition of water by Lavoisier in 1785, the 'fuel cell' has been not only theoretically possible, but also technically feasible. In England in the 1880s, researchers and industrialists tested generators and made precise calculations of their efficiency, and the research has continued uninterrupted.[28] But nothing could be done, because the fuel, i.e. hydrogen in the form of dihydrogen (H_2), had to be extracted from water. This gas is so light that it escapes from our atmosphere, where only minute traces remain. The electrolysis of water does provide the precious fuel, but the magic of electrolysis needs the electricity fairy, which in turn requires transformation from primary energy, so the problem is only displaced. Moreover, obtaining hydrogen by electrolysis of water is very unprofitable in practice (it is generally consumed on the spot to run the plants), and the only hydrogen production techniques with sufficient yields to allow a 'hydrogen economy' are those based on the combustion or cracking of fossil hydrocarbons. Two centuries later, water happily continues to fuel a free energy utopia, but doesn't run the engines.

Marcellin Berthelot, the chemist, was as much of a dreamer as Verne's hero, promising a bright future through chemistry. At a banquet in 1894, he predicted the end of agriculture, as well as of coal mining, by the year 2000:

In those times, there will be no more agriculture, no more shepherds, no more labourers in the world: the problem of maintaining existence by cultivating the soil will have been eliminated by chemistry! There will be no more coal mines, no more underground industries, and consequently no more miners' strikes! The problem of fuels will have been eliminated by the combination of chemistry and physics.[29]

At first glance, these past futuristic dreams all look the same. Yet their interest lies in their subtle differences. Verne's prose, steeped in the dominant model of the heat engine evoked in the beautiful image of burning water, sees no alternative; we will continue to burn the earth's resources. 'Water will be the coal of the future.' For Berthelot, on the other hand, the bright future does not consist in solving 'the fuel problem' by finding some kind of substitute for coal, but in 'eliminating' it with the help of physics and chemistry. The rest of his remarks do not shed much light on how to make this radical change, but at least he foresaw another model.

Just as burning coal was not an option in the eighteenth-century *oeconomic* perspective, to keep on burning the earth's resources by finding a substitute for coal was a choice that could have been rejected. But we persevered in our 'choice of fire'.[30]

11

The bewitching power of oil

Jevons had alerted the British government about the limitations of the primary resource on which the Empire's global dominance rested. But in 1865 he still believed there was no substitute to rival coal's performance:

> Coal in truth stands not beside, but entirely above, all other commodities. It is the material source of the energy of the country – the universal aid – the factor in everything we do. . . . *But the new applications of coal are of an unlimited character.* In the command of energy, molecular and mechanical, we have the key to all the infinite varieties of change in place and kind of which nature is capable.[1]

Yet in 1914, as head of the British Admiralty, Winston Churchill decided to equip the Royal Navy's ships with petroleum engines. Such a decision was somewhat surprising when coal mining in England was breaking records in 1913, and with a very, very optimistic inventory of reserves.

The rush for black gold

There were strategic reasons relating to Churchill's decision. As Timothy Mitchell notes, by 1914 oil had already conquered markets and built empires.[2] Drilling and exploitation had multiplied, financed by powerful bankers and investors. In Baku on the Caspian Sea, Robert Nobel, Alfred's brother, was pumping from three hundred wells in 1880. In the United States, Ohio, Pennsylvania and New Jersey were the oil-producing areas where John D. Rockefeller founded the Standard Oil Company in 1870. As the owners of the land were also the owners of the subsoil, Standard Oil initially focused on refining and marketing oil from thousands of domestic producers. Within a few decades, the company had vertically integrated the entire chain from extraction to refining and marketing.

On the European side, however, the rush for black gold was directed towards distant lands. Even though some oil was being exploited in Austrian Galicia and Romania, companies were prospecting for oil far away: the Dutch in Sumatra with Royal Dutch Petroleum, which was to become Shell, and the British in Persia, where the Anglo-Persian Oil Company (APOC), which was to become British Petroleum (BP), was created in 1908. APOC was more fragile because it operated in a more hostile environment than Standard Oil, so it sought allies and turned to the British Empire.

Churchill was all the more receptive to oil's appeal because, as Home Secretary in 1911, he had to deal with miners' strikes on British soil. The concentration of workers in coal mines and the transport of the mineral by barge and then by rail encouraged blockades to obtain social rights. Concerned by the power of the unions to paralyse the whole country, Churchill chose to make the Royal Navy independent of such power. By playing the oil card rather than relying on domestic resources, he anticipated the strategic role that oil could play in wartime. Not only did oil weigh less on board a ship than coal, give off less smoke and require fewer workers in the engine room, but it was also vital for transport, which in turn allowed it to be carried faster and cheaper over long distances. Shortly after the German and British navies converted to oil, the first fighter planes appeared in the skies, powered by an internal combustion engine fuelled by oil. The synergy between oil and transport catalysed the military industry.

At the same time, the growing civilian demand for oil was causing tensions. Shortly before the war, the car had already converted to the internal combustion engine, and thus to petrol. In the United States, Henry Ford sold two hundred thousand units of the famous Ford T in 1913. As highlighted by the famous episode of the Marne taxis requisitioned by the French army in September 1916, or the small X-ray cars equipped by Marie Curie, cars, which, like planes and ships, were powered by petrol, operated on all fronts during the Great War. Matthieu Auzanneau, author of a richly documented history of oil, reports these words from Georges Clemenceau in 1917: 'If the allies don't want to lose the war, France must fight in the hour of supreme Germanic confrontation and must possess the gasoline that will be as necessary as blood in tomorrow's battles.'[3]

A capitalist sorcerer

But wasn't capitalism the biggest winner of the black gold rush? Oil was a relatively secure source of profit that fuelled a new capitalism, more international, mobile and even liquid than the capitalism based on coal. A more speculative and more financial form of capitalism. States used to control their own coal mines, but oil slipped through their fingers. With oil, economic power shifted from industrial states to private companies. It is true that Standard Oil, which had a monopoly on all oil in the United States, fell foul of the anti-trust law in 1909. But its break-up into thirty-four companies changed almost nothing, as it was accompanied by the creation of the US Federal Bank by a consortium of private banks, all linked to oil. In short, state power was eroded in favour of large industrial groups. States often intervened to secure production zones in distant regions such as the Middle East by installing governments with whom they had made agreements. For example, in 1921, the British Crown created the Kingdom of Iraq after crushing a revolt by various local factions. In 1945, Franklin Roosevelt, on his return from Yalta, met King Ibn Saud on the battleship *Quincy* in the Red Sea and concluded an agreement: the United States undertook to protect Saudi Arabia, which in return undertook to supply the United States, Europe and Japan, which needed energy, and hence oil, to rebuild their economies.

Did states manage to regain control of oil? Nationalizations in oil-producing countries such as Bolivia in 1937, Mexico in 1938 and Iran in 1952 did not really succeed in curbing the profits of the oil companies. After three years of attempts to compromise with the Iranian Prime Minister, in 1953 the CIA fomented a coup d'état that reinstalled the Shah and allowed oil profits to be shared with foreign companies. The fifty-fifty solution, first imposed by Venezuela in 1948, did not weaken the oil companies either. Thus, at the end of the negotiations between the American company Aramco and Saudi Arabia in 1950, the 'golden gimmick' consisted of Aramco paying an oil tax to Saudi Arabia but no longer paying any to the US federal government. Thus, as Auzanneau points out, it was the US taxpayer who subsidized Aramco in Saudi Arabia.[4] The White House favoured American oil companies and, at the same time, prevented the producing countries from falling into the orbit of the Soviets. A capitalist liberalism triumphed, accompanied by an ostensible patronage, of which Rockefeller was one of the brilliant initiators.[5] On the eve of the 1973 oil crisis, seven companies – Jersey Standard, Mobil, SoCal, Texaco, Gulf Oil, BP and Shell – controlled 91% of Middle Eastern oil production.

The balance of power in favour of the big oil companies was so obvious, brazen and even offensive that, throughout the twentieth century, stories about oil were full of plots, schemes and machinations. How could one not paint everything in black when reading about the CIA's manoeuvres in the Middle East and Indonesia, the intrigues of Françafrique in Gabon or the trade of arms and oil that laid the grounds for terrorism? Even the oil shortage and the 1973 oil shock seemed to have been fabricated to drive up oil prices, at a time when the Organization of Petroleum Exporting Countries (OPEC) was demanding its slice of the cake.[6] In any case, the US did not discourage OPEC from raising prices.

Just as coal unleashed a drunken Promethean power, oil seemed to unleash the lust for winning, profit at any price; it fed a soulless ultra-liberal individualism. This individualism was embodied and celebrated in the success of the individual mobile car, which was supposed to give everyone the freedom to move around as they pleased. There were many factors that enabled the internal combustion engine to win out over the electric vehicle in the twentieth century and created such a path dependence[7] that this choice now seems irreversible.[8] Private companies and governments seem to have cooperated to create a worldwide addiction to this object of fantasy and desire.[9]

A gift from the earth

Is oil a hostage or an accomplice in these deplorable machinations? It must be admitted that its physical and chemical properties make it attractive and tempting for many adventurers. 'Nature's petroleum oil and natural gas are the greatest gifts we will ever have,' wrote George Olah, a carbon chemist and 1991 Nobel Prize winner.[10] This oil extracted by no one other than earth itself from stone (from which its name derives: *petra* = stone + *oleum* = oil) is rightly nicknamed 'black gold'. It is worth its weight in gold, or should be. In any case, it indirectly took the place of gold when the US dollar was detached from the gold standard in 1971, to be aligned with the price of a barrel of oil, and to serve as a reference currency in international trade. Of course, nothing predisposed hydrocarbons to serve the interests of the few private companies that dominate the geopolitics of the twentieth century. However, they have played a strategic role since antiquity:

> From the first steps of civilization, of agriculture, and of maritime trade in Mesopotamia, the drive to control hydrocarbons – like the need to secure water access – was one of the major causes of war, because bitumen was necessary for waterproofing irrigation canals and boats.[11]

The development of both merchant and military seafaring depended on the condensation and thickening of this viscous liquid. The Christian empire of Byzantium used oil as an incendiary weapon to repel the Arabs in the sixth century. The secret of this 'Greek fire', which even ignited the sea, was jealously guarded. And many centuries later, oil was used to make the napalm bombs used by the Americans in Vietnam.

However, it is above all as a fuel associated with the internal combustion engine that oil demonstrates its power. Its calorific value is double if not triple that of coal. Overall, the calorific value of fossil fuels depends on the carbon content, relative to hydrogen or water. One tonne of coal equivalent (TCE) equals 0.67 tonnes of oil equivalent (TOE). It must be added that this is for good-quality coal, such as anthracite, because for brown coal the equivalence is 0.33 TOE. Moreover, oil is less harmful to the environment than coal, which emits sulphur dioxide, nitrogen oxide, particles, heavy metals such as mercury and arsenic, and has a lower CO_2 emission rate (Table 1). The superiority of oil is now well recognized. But although its efficiency is still lower than that of natural gas, it is much less polluting than its predecessor, coal.

Table 1. Comparison of calorific value and CO_2 emissions for different fossil fuels.

	Calorific value (kg/kg)	CO_2 emissions by mass (kg/kg)	CO_2 emissions related to calorific value (kg/GJ)
Natural gas	58,000 (GCV)*	2.75	47
Oil	41,900	3.12	74.5
Coal	35,100	3.67	104.6

* GCV = gross calorific value.

Moreover, unlike coal, oil is both fluid and light. Because it springs from the ground once drilled to a sufficient depth, it requires few workers and is less laborious to extract than coal. This makes oil look easy and convenient. It can be transported in pipelines over long distances without the need for loading and unloading. Whereas coal had become a weapon of struggle in the hands of the miners, oil worked against them: 'In fact,' writes Mitchell, 'oil pipelines were invented … to circumvent the wage demands of the teamsters who transported barrels of oil to the rail depot in horse-drawn wagons.'[12] Only one oil strike – the Baku strike of 1905, which figures as a dress rehearsal for the Russian Revolution – is remembered by workers. Since the 1970s, oil has been criss-crossing the seas on supertankers that head to countries, at the traders' whim, where the price of a barrel of crude is the most attractive. As for the oil spills, which have recurred since the *Amoco Cadiz* spilled 1.6 million barrels of crude oil on the coast of Finistère in 1978, we continue to see them as simple accidents that do not alter the course of things.

Oil is at the centre of great productivist narratives, like the boom period of France's '*Trente Glorieuses*' (1945–75), which carefully dismiss alternative energies.[13] This 'stone oil' has entranced people for generations and so permeated our ways of life that there is little chance that the 'capitalist sorcery' fuelled by oil will cease any time soon.[14]

While oil's lightness and fluidity make it seem easier to process than coal, it still requires a series of cumbersome processes – petroleum refining – to make it usable. However, this disadvantage has been turned into an advantage, as the processing produces a number of hydrocarbon by-products that can be monetized when an outlet is found for them. The first step is to distil the crude oil by heating it to 370°C to recover butane, propane, naphtha, paraffin, diesel, heating oil and residues at the bottom of the column that can be re-distilled to produce lubricants and bitumen. Part of Rockefeller's fortune came from the way Standard Oil made use of all the by-products of distillation. One of them, kerosene, called 'lamp oil' or 'paraffin' at the time, replaced whale oil for street lighting in Boston and New York in the 1890s. Naphtha, one of the distillation products, can in turn generate a host of useful by-products through two other operations.

Cracking, a method invented in 1891 by a Russian engineer, Vladimir Shukhov, enables long carbon chains (5 to 12 CH) to be split into smaller, more volatile units. Thermal cracking, the first method used, gave products of mediocre quality. But catalytic cracking, patented by Eugène Houdry in 1936, makes it possible to work at lower temperatures while the catalyst – often a zeolite – can be partially regenerated. A product known as 'heavy diesel' can thus be transformed into a component of petrol for cars. This cracking technique produces unsaturated hydro-carbons that can be used directly to manufacture mainly synthetic fibres and plastics, but also medicines, fertilizers, detergents, lubricants, solvents, waxes, pesticides and herbicides. Reforming, a technique requiring heating, high pressure and a catalyst such as platinum, aims to rearrange linear chains into rings to obtain aromatic hydrocarbons (benzene, methylbenzene, dimethylbenzene, ethylbenzene), all of which are basic products of the petrochemical industry. Alongside the refineries are large petrochemical complexes where molecules derived from oil are isomerized and polymerized to obtain useful chemicals. The BASF site in Ludwigshafen on the Rhine stretches for almost ten kilometres.

Virtues as traps

Despite the cumbersome processing carried out in refineries and large petrochemical complexes, oil has fed a whole mythology of ease, abundance and growth. Figure 12, which, like a medieval *mappa mundi*, shows the known territories of the oil world in 1946, makes a case for this. In this map, Harold Smith, a researcher at the Oklahoma Station Bureau of Mines, depicted petroleum products as rivers flowing from a carbon and hydrogen summit situated, laughingly, in the strato-sphere. The different petroleum products, distributed according to their chemical composition, are arranged concentrically according to their boiling point.

Black gold seems to be flowing in waves to irrigate the chemical industry, inscribing the paths of modernity and prosperity. We are caught in the grip of the oil world. Forty years of mobilization and environ-mental summits have hardly improved the chances of getting a grip on this sprawling world. Yet the age of fossil fuels now seems to be closing in. The idea of making petrol from coal by adding the Fischer–Tropsch process, as Germany did in wartime, has been abandoned, although the reserve-to-production (R/P) ratio of coal is three times that of natural

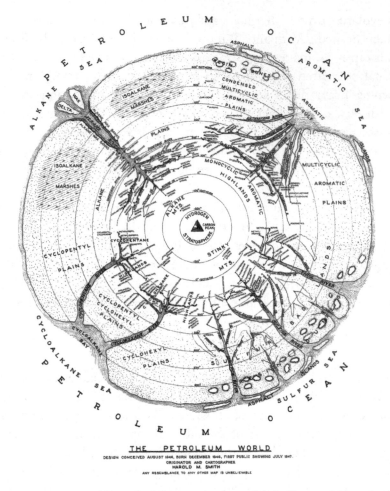

Figure 12. Harold Smith, 'The petroleum world' © Chemical Heritage
Foundation, Philadelphia.

gas and four times that of oil. Instead the focus is now on unconventional oils such as shale oil and light tight oil, which are being exploited particularly in the United States and Canada. While the estimation of oil reserves remains a matter of debate, it is even more uncertain for these non-conventional oils because there is no deposit concentrated in one locality. These oils, which are trapped in the bedrock, are spread over millions of square kilometres and can only be released under high pressure or by fracking techniques.

Even more adventurous are plans to exploit another fossil energy source: clathrates, or methane hydrates. These are molecular pockets of very pure methane trapped in crystal ice that form on the seabed. The large deposits discovered in the 1960s in the permafrost of northern Russia are attracting particular attention from the Chinese, Japanese and Koreans. But they are not about to revolutionize the supply of fossil fuels as they are very difficult to exploit. They are called 'flammable ice' because these thin cages of ice are only stable at low temperatures (around 0°C) and high pressure. Moreover, any accidental release of large quantities of methane into the environment would cause environmental disasters that we scarcely dare imagine.

The scenarios for the evolution in the demand for fossil fuels are just as uncertain. They take into account demographic developments, economic growth, prices, policies to combat the greenhouse effect, technological developments and many other factors that are not always predictable.

But whatever the success of alternative energies, it seems that oil will remain king in the transport sector for another twenty or thirty years despite the determination shown by some governments to speed up the energy transition in the private car market. Although electric cars allow the reduction of national carbon emissions of highly nuclear-powered countries, the manufacture of their batteries has still a high carbon footprint.[15] And even if we succeed in making the shift to the hydrogen economy that was hoped for at the beginning of the twentieth century thanks to fuel cells, can we really talk about 'non-carbon' energy? Hydrogen is difficult to store and transport. Methanol, which certainly does not produce energy, could be useful for storing energy in the form of hydrogen and thus become the companion or commensal of fuel cells. In short, carbon would still be used.

Thanks to its 'virtues', oil has been influencing world politics for over a century. Whereas coal favoured a territorialized technical system with local actors and power relations, oil has favoured the constitution of open networks and global powers without assignable territory. This means that both coal and oil have the power to shape the field of possible political futures in human societies. They clearly demonstrate that normativity is not the exclusive prerogative of humans and that material things also have the power to impose norms, to create alliances, to change customs and habits. Coal and oil have reorganized the way people live, think, consume and trade in most countries of the world. The need to extract, transform and circulate carbon involves alliances between natural, social, geographical and political elements that are very different in the case of coal and oil. Although they may have unleashed an intoxication for all things technological and dreams of a dematerialized economy, these two carbons with such contrasting physical properties tell us how intimately human adventures are associated with the 'virtues' of materials.

What does this metaphorical use of the term 'virtue' mean? Both coal and oil have a *virtus*, a moral force in the sense that they convey norms and values. Coal has spread values such as technical, economic and social progress. It has contributed as much to the imposition of productivist norms as to the expansion of social rights and the regulation of working conditions from the mines. Oil, in turn, has spread individualistic and liberal values by adding consumerist norms to the norms of productivist growth. Both have fuelled the modern ideal of emancipation from the constraints of nature, spatial distance and time. But the power of coal and oil comes from a blindness as to their ontological status. People wanted to see these carbon compounds as simple reservoirs of potential energy to be released. The chemical bonds between carbon and hydrogen have been 'cracked', and carbon has been consumed as if it were food, only to be abandoned in favour of hydrogen again. In doing so, there was an intention to neglect the fact that these molecular architectures were works of time, that they were formed over the long duration of the earth's history, over hundreds of millions of years. As we have seen above,[16] all fossil energy is concentrated time. To consume it is to consume time. Burning a tank of petrol to go on holiday or to fly from Paris to New York is burning a few thousand years in a few hours. Thus, economies based on coal and oil have established a regime of time-consuming temporality. The paradox is that the more time we consume, the more time we lack. The imperatives of growth, with productivist and consumerist norms, result in an acceleration of technical innovations that generate a chronic feeling of time shortage, as the German critical theorist Hartmut Rosa has pointed out.[17] The prevailing impression of acceleration[18] means that the promises of emancipation inscribed in modernity are, on the contrary, met with a narrowing of temporal frameworks, a feeling of 'not enough time', of doing things in a hurry and a fixation on the short term. Rosa's diagnosis is clear: it is time that is in crisis.

But is it not rather a matter of the relationship of modern societies with carbon compounds? For more than two centuries, molecular architectures that have been skilfully crafted over millions of years in the bowels of the earth have been burnt and released in smoke. We assigned to carbon only a consumption value at the expense of its technical qualities, its cultural values, not to mention its ecological value.

12

The age of plastics

'Plastic happens; that is all we need to know on earth.'[1] This line from Richard Powers' novel *Gains* sums up the situation in a few words. The novel tells the story of how a small family-owned candle and soap factory in nineteenth-century Illinois grows into a powerful company – like Procter & Gamble – that floods the national and international market with chemicals of all kinds. '[T]he company's molecular toolbox,' writes Powers, 'ought to be producing a new product once every forty-seven days.'[2]

Although plastics were not part of the catalogue of products manufactured by this factory, which specialized in soaps, shampoos, fertilizers and detergents, the author presents the disposable plastic camera as an emblem of synthetic production to hammer home his message: chemical companies mobilize the full potential of an arsenal of molecules in order to make a profit, a surplus value, with no regard for what these molecules do to consumers, users and the world. Indeed, as a counterpoint to the rise of industrial capitalism, this novel unfolds the personal drama of a local woman suffering from ovarian cancer, probably due to the chemicals released into the environment. To her ex-husband, who tries to drag her into a lawsuit against the chemical company, she replies that there is no point in exhausting her strength in a lawsuit because it is impossible to quantify the damage caused by the company. 'Plastic happens.' It is part of our lives, part of our world.

What is at stake is not only the powerlessness of individuals in the face of the great capitalist corporations, but also the mode of existence of the thousands of molecules synthesized to satisfy needs that are just as artificial as they are. These plastics, once they come into existence, inscribe their signatures in the world, because they bring about material transformations.

So, we have a new life to reveal for carbon, its trials and tribulations in the world of mass consumption, intoxicated with oil.

'Better things for better living . . . through chemistry'

'Better things for better living . . . through chemistry.' This famous slogan came out of the advertising department of the DuPont firm, which had built its reputation on the production of explosives in its Wilmington factory near Philadelphia. The slogan, invented in 1935, was used until the 1990s, with the 'through chemistry' edited out in 1982. By insisting on 'better living', its main aim was to make people forget the images of death associated with DuPont products since the war and to promote synthetic products for everyday life.

In the 1920s and 1930s, at the instigation of its director of chemistry, Charles Stine, DuPont invested heavily in research and recruited several hundred chemists to redeploy its activity in the manufacture of synthetic polymers.[3] In 1928, a Harvard chemist, Wallace H. Carothers, was appointed to lead the research into long-chain polymers. In 1931, DuPont began manufacturing synthetic rubber with a product simply called Duprene. The synthetic rubber manufactured by DuPont was derived from the identical repetition of a monomer – chloroprene, C_4H_5Cl – whose semi-developed formula is $CH_2 = C(Cl)CH = CH_2$. It is like a 'disciplined' latex due to the substitution of the methyl radical of isoprene with chlorine, an oxidant that blocks 'anarchic' (natural) association potentials to force the chains of the carbon skeleton into a neat alignment.

Duprene's commercial potential was limited by its foul odour, so a new manufacturing process was developed that eliminated the unwanted smell. The material then got a foothold in the market for waterproof gloves and clothing. To prevent other manufacturers from damaging the product's reputation, the Duprene brand was limited to products made by DuPont. But as the company abandoned the manufacture of finished products containing Duprene, the material was renamed neoprene in 1937, a name that emphasized its status as a generic ingredient, not a finished consumer product. The first wetsuits created by Jack O'Neill were made of neoprene. They also inspired the fashion for tight-fitting, brightly coloured jumpsuits for Hollywood starlets.

In 1934, just as Lammot DuPont and his brothers were being publicly accused of taking advantage of the war to enrich themselves,[4] Wallace H. Carothers synthesized a new polymer that would change the future of the company.[5] Julian W. Hill, his assistant, said that when he saw a whitish, filamentous paste emerge from a test tube, he 'felt the molecules fall into place in parallel lines and the hydrogen atoms cling to each other'.[6] This paste, formed by the polycondensation of a dibasic acid[7] and a dry diamine,[8] hardens in the air into an elastic fibre that can be stretched. It is called polyamide 6-6 because each of the constituent molecules of hexa-methylene diamine and adipic acid has six carbon atoms. It was first commercialized in 1938 in a toothbrush in which pig bristles were replaced by the synthetic fibre, protected by a patent in which it was named Nylon.

Since the nineteenth century, many chemists have looked for substitutes for animal and vegetable fibres and natural pigments. The first attempts tried to imitate the silkworm, which digests the cellulose in mulberry leaves and produces the fibre that made China prosperous for almost four thousand years. To 'digest' the cellulose, chemists treated it with nitric acid, which was not ideal because of the risk of explosions. Nitrocellulose was used to make powder like celluloid. Nevertheless, the process enabled Count Louis-Marie Hilaire de Chardonnet to patent the first artificial silk in 1885 and to market it successfully. But this artificial silk was soon dubbed 'mother-in-law silk' because of the fires and explosions that occurred in several factories.[9] In Britain, Charles F. Cross, Edward J. Bevan and Clayton Beadle had better luck treating wood pulp with caustic soda and a few other products to obtain a slightly golden substance which they called 'viscose'. The process enabled the English textile manufacturer Samuel Courtauld & Co., a world specialist in the manufacture of crêpe for mourning, to move out of the mortuary environment and into the manufacture of rayon at the beginning of the twentieth century, and then acetate in the 1920s. With these semi-synthetic fibres – still made from wood cellulose – Courtauld conquered the hosiery and lingerie markets in Europe and the United States. Viscose and acetate gave rise to a new aesthetic, a new fashion and even a new world, which *Fortune* magazine put on the map as a new continent called Synthetica (Figure 13). This 'new continent of plastic' emerges from natural materials, represented as nations: resins, lignin, oil, cellulose. The farther the continent moves into the ocean, the farther it moves away from natural land. Cellulose covers an important but not very advanced territory. It is from Petrolia

Figure 13. The synthetic continent created by Ortho Plastic Novelties, published in *Fortune* magazine, July 1940.

that the new nations are developing as the by-products of cracking oil for fuel production provide plastics with a raw material that seems almost free. The still emerging Nylon country is represented by an island. The carbon, hydrogen, oxygen and nitrogen that go into the formation of these molecules are shown as the four cardinal points that serve as landmarks.

Firmly established in this synthetic continent, DuPont was not content to imitate natural silk. It sought to synthesize long, soft, strong, elastic fibres that could be used to make both silk and camera film. A versatile, multi-purpose plastic fibre was the surest way to ensure a return on the heavy research investments made by the firm. This bet on generic plasticity was in line with a logic of economic profitability but also with the material logic of carbon, which is used in the composition of a host of materials thanks to its potential for combination and structural innovation. This gamble is also reflected in the trade names given by DuPont to polyamide 6-6 and polychloroprene – nylon and neoprene rather than artificial silk and synthetic rubber. The wager on plasticity was not necessarily a good one, however, judging by this comment in the press:

Man, after experimenting for years … has finally discovered that by an ingenious mixture of castor oil, ethylene, glycol, carbon, hydrogen and oxygen, he can make a silk fibre almost as good and not more than three times as expensive as the one the Chinese worm has been manufacturing for centuries.[10]

Despite such remarks, and thanks to the thousands of dollars invested in research and development, in chemical engineering, as well as in advertising campaigns on billboards, on the radio, or in exhibitions, nylon dethroned natural silk.[11] On 15 May 1940, a big national sale of nylon stockings announced by radio attracted thousands of women, who jostled each other at the department stores. After a pause during the war years, when nylon production was redirected to military applications, DuPont relaunched mass production in 1945 with spectacular promotional sales. There was a nylon rush, with long queues on the pavement, a riotous atmosphere, hysteria. The brand name Nylon, unprotected by the firm, quickly became a household name. In 1945, DuPont sold sixty-four million pairs of nylon stockings at twice the price of natural silk stockings. In a period of tension between the United States and Japan, the press interpreted this commercial success as a boycott of Japanese silk. Nylon, in any case, had conquered American women and earned DuPont $25 billion for an investment of $4.3 million.[12]

Not content with its orchestrating of a grand stage show to promote nylon stockings, DuPont stepped up the pressure to repeat the nylon miracle every year. The firm developed a kind of religion of innovation at all costs, backed by heavy investment in research, by flooding the market with new man-made fibres. By the end of the 1940s, DuPont was producing and marketing more than a thousand different varieties of nylon and new fibres – polyester, acetate, acrylic, etc.

The miracles of plastic

Advertising campaigns conjured with the miraculous, with magic. Chemistry would transform base materials into high-quality fabrics. Synthetic fibres made luxury available to all and their mass distribution imparted an aura of democratization. They combated economic depression and ushered in an era of abundance. They preserve nature by sparing rubber trees and animals that provide hair, fur or ivory. Yet synthetic polymers did not come into the world as simple substitutes, mere ersatz materials. While the first plastics – such as celluloid – were disappointing in the way they imitated natural materials such as wood and tortoiseshell because they felt a bit fake and cheap, nylon founded a new culture, a cult of synthetics.[13]

Synthetic polymers are grouped under the generic label 'plastics'. Before it was made substantive, this term qualified both objects and subjects, as any instance that both gives and receives form. The term is derived from the Greek verb *plassein*, which means to shape or to mould. It can take on a figurative meaning of feigning, imagining or lying when applied subjectively.[14] In such cases, however, it is an essentially positive attribute which designates, in neurology as in psychoanalysis, the way in which the organism is formed under the effect of experience, as Catherine Malabou argues.[15] This meaning predominates in a 1925 film entitled *The Plastic Age*, which features teenagers with malleable and changeable personalities. But the term was soon taken over by carbon chemistry. As early as 1932, a book entitled *Chemistry Triumphant* announced a 'silico-plastic age'.[16] The lightness and adaptability of materials that could be used to make all sorts of objects no longer connoted a lack of authenticity, but rather flexibility and adaptability, two rising values in twentieth-century culture. As Jeffrey Meikle shows, plastics were more or less in line with the American way of life of the 1950s, in that they are versatile, they adapt to all uses, to all functions, and they are as fashionable as they are domesticated.[17] Disposability is the secret of the commercial success of plastics, allowing for the mass production of objects intended for a single use.[18] And the ephemeral life of all these products was far from being a defect; it was valued as a guarantee of hygiene. It welcomed change, with the proliferation of objects saturating domestic spaces and creating a new aesthetic. The modern woman changed her wardrobe, her jewellery, her cosmetics, her hair: a 'plastic princess' evolving in a consumerist paradise.[19]

By lending themselves to all desires and uses, to all objects, from the most common and everyday to works of art, carbon chains came to become pure metamorphic capacities. But do these carbon chains, which can take and give all possible forms, really have a proper nature? As Roland Barthes clearly saw in his *Mythologies*, plastics dismantle the category of substance to give way to a metaphor of flow, of becoming. 'So, more than a substance, plastic is the very idea of infinite transformation. . . . It is less a thing than a trace of a movement.'[20] The chemical process that transforms the few initial crystals into fabric, a bowl or a piece of jewellery seems to be frozen for a moment, like a freeze frame in a film. This is the 'miracle' (in the sense of a break in nature) of polymerization, which ensures that the material is set and shaped at the same time.

This status of quasi-virtuality, at the limits of materiality, confers on plastic objects a raft of values – freedom, lightness, flexibility, adaptability. All of these values are held high in a society of 'plastic dreams',[21] which values superficiality in the artificial. Hence the deep contradictions that are attached to plastics. Their promises of indefinite change are like the stuff of dreams, but they lack identity, personality and depth. Moreover, it is by virtue of the possibility of indefinite reproducibility that plastics become objects of desire.[22] Paradoxically, the repetition of the same is supposed to fulfil individual desires. Finally, there is a striking contrast between the lightness of plastic objects, which allows for triviality and frivolity, and the heaviness of the technological chains that extend from oil wells to the boats and containers that transport oil to the chemical factories with their extrusion, moulding and bottling equipment, and to the trucks that transport the finished plastic goods to the shelves of supermarkets and into the household shopping bag. The fact remains that this 'disgraced material', as Barthes puts it, has a strong normative power: it commands change, single use, disposability, waste bins. 'Plastic is wholly swallowed up in the fact of being used.'[23] Yet, since the 1970s, the uses of plastic materials have continued to expand.

Reinforced with carbon

It is important to note that plastic's virtues do not come from its polymer chains alone. Many additives are incorporated to facilitate polymerization, as well as to optimize the end product; to reduce the price, improve mechanical strength and thermal stability or reduce flammability and delay ageing.[24] In fact, plastics need expert work in 'formulating' resins, an art of mixing which has enabled the market for synthetic polymers to extend far beyond women's stockings and underwear to aircraft wings. This is because the polymer chains that have already realized the potential for combining carbon atoms are also involved in blending with various ingredients. The resin becomes a matrix that accommodates reinforcements and distributes mechanical stresses, increasing compressive strength. This is the principle of composites, which always combine a matrix and a reinforcement, a continuous medium and discrete elements, two phases, regardless of the nature of the components.[25] Combining the lightness of plastics with the hardness of steel and the strength of ceramics, this dream for materials engineers became a reality in the context of the major space and military programmes of the 1960s, which financed cutting-edge research with no hope of an immediate return on investment as they aimed for high-end multifunctional materials. To manufacture such chimeras, carbon fibres are very attractive because of their high elastic modulus. In the competition among materials, carbon offers good structural efficiency due to its low density and stiffness. When embedded in epoxy or polyamide resins, carbon fibres perform five or six times better than aluminium or titanium alloys.[26]

These fibres renewed composite technology in the 1980s. Traditional moulding processes, which involve the production of semi-finished products such as fabrics, sheets or films which are then polymerized, were replaced by reinforced reaction injection moulding (RRIM), which directly shapes parts from the monomer and the fibres. Hollow shapes such as cones, cylinders or spirals can be moulded, depending on the final object. The carbon fibre gives the carbon chains additional plasticity. In addition, it is conducive for the production of anisotropic materials, with a special topology adapted to the desired functions for a specific part. Aircraft manufacturers have learned to design landscape materials with fibres distributed and oriented in the resin according to the stresses exerted on the part to be built. A mesh or fabric of fibres is designed layer by layer and the topology of the material is thus built up by successive layers to obtain a customized material.

This marriage of resin and fibre means we have to look again at the materiality of the components. Indeed, the quality of the material depends on the bond between matrix and reinforcement. Whatever the modulus of a carbon fibre, if it does not adhere well to the resin, the material will fail. To obtain a composite that holds together, the fibres must be pre-impregnated. Hence the development of surface chemistry, an art of managing interfaces, a know-how of relationships that doubles the jobs done by the bonds and the relationships of carbon atoms.[27]

The process of manufacturing carbon fibres from PAN (polyacrylonitrile $[CH_2-CH-CN]_n$) is complex and energy-intensive: oxidation, then carbonization and finally graphitization to obtain a 99% carbon fibre.[28] In terms of cost, carbon fibres are less efficient than other reinforcing fibres. Because of its cost price (an average of 2,000 francs per kilo in the 1980s [about 842 euros today]), 'black fibre' has not supplanted white fibreglass on the consumer product market and remains confined to a niche of high-performance materials. Thanks to carbon fibre, composites are replacing aluminium bit by bit because of their lower density (1.6 compared to 2.8 for aluminium). In 1973, Concorde used them only for the fairings and wingtips. But gradually, because of their length (greater than that of competing aramid fibres such as Kevlar), carbon fibres have managed to spread to the mass production market thanks to sporting goods such as windsurf boards, skis, tennis rackets, etc. Extreme sports, new sporting achievements and records – all have greatly benefited from carbon fibre.

Once again, carbon is shaping our lifestyles. While cheap plastics have spread the culture of disposability, carbon fibre composites have fostered a cult of performance in society. Synthetic polymers had already inspired a culture of artifice. They had, according to French philosopher François Dagognet, 'definitively severed the ties that still linked us to the earth, the grasses and the meadows'.[29] Concerning composites, as a result of space conquests, Dagognet speaks of 'defeating nature' and 'the battle that was played out between nature and the laboratory'.[30] Yet, even though they invite technical prowess and seem to distance us from nature, paradoxically, carbon fibre composites lead us back to nature. In the 1980s, they redirected designers' attention to natural materials, where the fibres also play a structural role, depending on their orientation. Many materials researchers have collaborated with marine biologists to study the fine structure of shells, such as abalone. Biomimicry is now a strong trend in materials science and engineering, and is making steady industrial progress.[31]

If carbon's fibrous state somewhat tempers the 'societal impact' of its resinous existence at the origin of disposable fashions, the union of fibres and resin in the composites worsens its 'environmental impact'. Composites certainly allow for manufacturing savings because several functions can be integrated into a single moulded part instead of assembling the various functional elements one by one with screws.[32] However, this advantage is paid for in terms of use, as, in the event of a breakdown or breakage, the entire part must be replaced. The savings in manufacturing therefore translate into increased maintenance costs. As for recycling, it is even more difficult with materials made from subtle mixtures.

A continent of waste

Less than a century after the triumph of nylon over silk made by worms, the virtues of nature are being rediscovered. As soon as innovation takes life cycles into account and not just performance and price, nature becomes a model. Where does this craze for natural materials, biopolymers and biocomposites come from? Not only do natural materials offer enviable performance, but above all their recycling is almost built into their design. Yet plastics recycling has become a major issue in recent decades.[33] Even though they are derived from fossil carbon, polyethylene plastic bags are not biodegradable. They are, however, photodegradable. Therefore, after a few years of exposure to the sun, they tear, crumble and break down into microscopic fragments.

In 1972, two oceanographers communicated to scientific journals the existence of significant concentrations of synthetic polymer particles in the Sargasso Sea, located between Florida and the northern Bahamas; a similar observation was made in 1973 for the North Pacific.[34] Scientific publications have restricted access, which is not enough to alert public opinion. Just as Rachel Carson had chosen in 1962 to alert the public to the pollution of the oceans by pesticides by means of her fable-like story *Silent Spring*, Curtis Ebbesmeyer, an American oceanographer and specialist in marine currents, used an allegory to raise the alarm. The metaphor of a continent of rubbish – the 'Eastern Pacific Garbage Patch' – was suggested to him by his sailing friend Charles Moore, who recounted his 'encounter' at sea with plastic debris floating all around him in isolated waters of the Pacific.[35] The metaphor has the advantage of underlining the scale of the phenomenon. Of the three hundred million tonnes of plastic used each year, eight million end up in the oceans.[36]

This is a mirror image of the Synthetica continent celebrated in 1940 as the new found land of technology. However, the so-called 'continental island' of waste does not have the solidity of a continent, but exists rather as a gyre or whirlpool, as aptly suggested by the apologue with which Michel Serres opens his book devoted to the genesis of objects made up of multiplicities stemming from chance and noise.

> As I was sailing along that summer, under a dazzling sky, and drifting lazily in the wind and sun, I found myself, one fine morning, in the green and stagnant waters of the Sargasso sea. … I was dumbfounded to see an area almost two hundred and fifty acres square entirely populated by dancing bottles. There were countless little vessels. … The coiling winds had compelled them all here, from far and near, from a thousand different quadrants. There constant and perilous collisions made for an acute and cacophonic carillon.[37]

It is now assumed that there are five such gyres or eddies formed by subtropical current convection, and it is estimated that the North Pacific gyre occupies about 3.43 million square kilometres, or one third the size of Europe. Indeed, the metaphor of the 'seventh continent' persists and is even used as the banner for a whole series of expeditions supported by the Centre international d'études spatiales (CNES) and the European Space Agency (ESA).[38] The figure of a large, visible and tangible mass seems more likely to alert public opinion to the global ecological problem linked to our consumption habits than the metaphor of a 'soup' with a variable concentration of floating debris. However, the synthetic polymer chains will not aggregate to form a solid, compact mass once again. On the contrary, the bits of plastic disintegrate under the combined effect of the sun's rays and the abrasion of the waves until they become imperceptible; tiny fragments that cannot be collected or counted, to which are added plastic microbeads or nurdles known as 'mermaid's tears'. As Baptiste Monsaingeon points out, the 'plastic soup' image is much more worrying because it prevents us from dreaming of a major clean-up to rid ourselves of all this waste.[39] The elusive and microscopic nature of ocean plastics also casts doubt on bold recycling projects proposing to turn these pseudo-continents into real territories, islands of ecological habitats, through which we would redeem our 'plastic sins' (Figure 14).

'Plastic happens; that is all we need to know on earth.' Richard Powers' thought mentioned at the beginning of this chapter reminds us that there is no possible return to a state of original purity. The fragmentation process can go on until nanoparticles are produced that fish can eat like plankton and thus enter the food chain. In the meantime, millions of floating particles provide ideal niches for a host of micro-organisms or even spiders that happily colonize previously inaccessible territories. This has led to a profound transformation of marine ecosystems. The carbon chains that were skilfully put together to produce all sorts of materials with new properties are having a hard time. They are entering into a new mode of existence with bacteria and spiders which, unlike humans, know how to 'make worlds with' these carbon molecules.

As plastic bottles, one of the most successful plastic commodities, turned out to be a major source of plastic waste, it is worth concluding this chapter on Primo Levi's ironic remark about polyethylene. When he was a prisoner trying to survive in Auschwitz by stealing from the chemistry laboratory everything that could be exchanged for a piece of bread, he was desperately in need of a container for liquids. He wrote:

Figure 14. The *Recycled Island* project, proposed by the Dutch architectural firm WHIM in 2009 © WHIM architecture.

This is the great problem of packaging, which every experienced chemist knows: and it was well known to God Almighty, who solved it brilliantly, as he is wont to, with cellular membranes, eggshells, the multiple peel of oranges, and our own skins, because after all we too are liquids. Now, at that time, there did not exist polyethylene, which would have suited me perfectly since it is flexible, light, and splendidly impermeable: but it is also a bit too incorruptible, and not by chance God Almighty himself, although he is a master of polymerization, abstained from patenting it: He does not like incorruptible things.[40]

13

Working towards a more sustainable economy

Although it's fair to say that fossil energy comes from the sun, it materializes in carbon via living organisms. Fossil fuels are either hydrocarbons (C_mH_n), in the case of oil and natural gas, or carbohydrates ($C_m[H_2O]_n$), in the case of coal. In fossil fuels, as in living things, carbohydrates are sources of energy producing work. Hydrocarbons being used in industry and transport are linked to economic growth and national independence; carbohydrates are in the cells of living organisms and used for nutrition and health. Machines on one side, life on the other.

They seem to be quite at odds, at least with regard to their constitution in time. The linear time of economic expansion and growth associated with technologies contrasts with the cycles and loops of time in living things. Two carbon metabolisms are at play with their own asynchronous tempos: on the one hand, industrial metabolism, which is connected to the slow anabolic processes of hydrocarbon sedimentation and the rapid catabolism of their unbridled consumption;[1] on the other hand, biological metabolism, which combines rapid anabolism of organic carbon (feeding, respiration, photosynthesis) and slow catabolism, which is registered in generational and evolutionary unfolding, or, in a small way, in the sedimented time of soils.

A gulf thus seems to separate the world of coal or oil from the muscular strength of animals and humans, which has given way to machines powered by fire. But the source of energy is the same: it lies in the carbon chains produced by living organisms.

So, couldn't energy be derived from the primary source, living organisms? We might as well short-circuit the long sedimentation time of fossil fuels and connect industrial metabolism directly to the rapid cycles of living metabolism. To bring about what is modestly called the 'ecological transition', two types of measures are being put in place: on the one hand, a bioeconomy aimed at using living organisms directly as a source of energy and renewable products; and, on the other hand, a carbon market aimed at offsetting mineral carbon emissions by creating organic carbon.[2]

From black gold to green oil

For nearly a century, petrochemistry has been distilling, cracking, splitting and reforming carbon chains to produce the chemicals it wants with stable bonds and cycles. The processes used in petrochemistry may be expertly developed, calculated, rationalized and optimized, but they still *degrade* in the strict sense of the word, as molecules that have been patiently woven together by living organisms and endowed with multiple, intertwined and synergistic functionalities are broken down. We reduce these jewels of complexity into small purified molecules that must then be reformed or polymerized to make them function again. Wouldn't it be more logical to draw carbon chains from the source and put them to work directly?

The idea of drawing carbon chains directly from living things is certainly not new. In 1941, Henry Ford impressed the industrial world by showing the public the Soybean Car, a car whose plastic body contained ingredients such as soya, wheat, hemp and flax. This initiative reflected a concern for the integration of agriculture and industry, driven by a movement called 'Farm Chemurgic', which began in the United States in 1935. The chemist William J. Hale urged industry and governments to expand agricultural markets to reduce the industry's dependence on foreign countries.[3] During the Second World War, the Farm Chemurgic Council played a strategic national role in dealing with the shortage of raw materials. It made attempts to synthesize rubber from maize and sent schoolchildren into the fields to collect common milkweed, a wadding weed used in the manufacture of military lifejackets. But agricultural carbon remained a wartime substitute with no commercial future.

However, the idea was revived in the 1960s with the first alarms about the sustainability of fishing practices, and then of the economy in general. In this context, Nicholas Georgescu-Roegen developed an evolutionary and thermodynamic reading of the economic system which clearly warned of the cost of non-renewable energy and irreversible destruction.[4] Since it cannot free itself from thermodynamic and biological constraints, the economy must be part of ecological cycles.

Georgescu-Roegen thus advocated a bioeconomy that limits the waste of natural resources and the proliferation of waste, reduces the gap with underdeveloped countries and looks after future generations instead of seeking to maximize profits in the present. The 'entropic dowry of humanity' is such that:

> From the viewpoint of the extreme longrun, the terrestrial free energy is far scarcer than that received from the sun. The point exposes the foolishness of the victory cry that we can finally obtain protein from fossil fuels! Sane reason tells us to move in the opposite direction, to convert vegetable stuff into hydrocarbon fuel.[5]

Substituting renewable carbon for fossil carbon to produce energy and materials is one of the watchwords of bioeconomics. Hydrocarbons are being replaced by carbohydrates. The chemical farm will be the result of planning exercises that identify interesting platform chemicals and draw up roadmaps for reaching these targets, such as producing intermediate chemicals, based on sugar rather than oil, for example to produce ethylene from ethanol. The aim is to eventually replace industrial chemicals with biochemical production.

But has the petrochemical model of molecule degradation really been abandoned? The concept of the biorefinery, used to designate units for the production of biofuels from agriculture, bears witness to the influence of this model. 'Green chemistry' mimics petrochemistry, which is subdivided into a *basic chemistry*, designed to produce five major intermediates, and a fine chemistry, using these intermediates as precursors to develop its own products for a range of industries. Unsurprisingly, biorefineries are set up on the sites of chemical complexes. The logic remains one of cracking carbon chains and rebuilding them from purified basic elements. Hence the paradox highlighted by Roland Verhé, editor of the *European Journal of Lipid Science and Technology*: in petrochemistry, we start with a distillation product such as naphtha and add functionalities to this small carbon chain to obtain detergents, surfactants, lubricants, etc. In contrast, in biorefineries, we start with plant essences that already have long carbon chains with multiple functionalities and by 'fractionation of plant extract' we produce the desired chemicals. However, it is absurd from a chemical and economic point of view – at least in terms of saving atoms and energy – to destroy the functionalities present in the carbon chains of vegetable oils in order to produce biofuel. Would it not be more reasonable to explore and exploit their potential?[6]

The processing of 'green oil' calls for another contrasting model. As Martino Nieddu et al. point out, instead of splitting carbon chains to obtain basic elements, to obtain functionalities it is better to preserve the architecture of the long carbon molecules made by living organisms with their twelve to thirty carbons.[7] It would therefore be necessary to replace the logic of cracking and reforming with a logic of extraction, requiring delicate physicochemical treatment. For example, the treatment of starch to obtain a PVC substitute spares the carbon chains instead of destroying them to recompose them. This type of chemistry is rediscovering the art of extracting what eighteenth-century chemists referred to as the 'immediate principles'[8] of plants and is mobilizing the know-how of pharmacists and perfumers. It allows the manufacture of high value-added speciality chemicals in small units located on the sites of agricultural production, targeting a niche market rather than mass production and consumption.

Thus chemists who are rediscovering the value embedded in the rich carbon structures developed by plants during their evolution, rather than in the cracking of fossil oil, are learning to develop a bioeconomy in contrast to the petroeconomy.

Towards a white carbon?

Since the 2000s, another concept of the bioeconomy has tended to take precedence over the original concept.[9] The OECD report 'The Bioeconomy to 2030' provides a glimpse of the future of carbon in the global economy. Living matter is presented not only as the miracle solution for providing renewable energy without depleting fossil fuels, but also as the secret for growth and prosperity. So bioeconomy is referring to a world in which biotechnology makes up a significant proportion of economic production.[10]

Living things have certainly been serving the economy ever since the development of agriculture and animal husbandry. But OECD-style bioeconomy is not about grains or animals grazing in the fields. The first pillar of this new bioeconomy is in the laboratory more than in the field. It is about research and development of new pharmaceutical products (vectorized medicines or recombinant vaccines), new plant and animal varieties, industrial yeasts, etc. This knowledge comes from molecular biology, genetic engineering, genomics and proteomics, which mobilize bioinformatics.[11]

This bioeconomy is based on the use of the basic building blocks of life. The processes exploiting these carbonaceous molecules were called 'white biotechnologies' in the 2000s to attract investors to a promising field, quite distinct from the 'red biotechnologies' in health (focusing on antibiotics or antidepressants), as well as from the 'green biotechnologies' (diverting agriculture's aims from food for the benefit of industrial production). Biotechnology is at the forefront of this heralded new global economy. Although the aim is still to reconcile the economy and ecology, we can speak of biocapitalism insofar as it is no longer labour that is presented as the source of added value, as in the capitalism described by Marx, but life itself as the capacity to organize matter.[12] The advantage of using building blocks rather than carbon structures extracted from certain plants is that one can target generic products and production capacities and no longer just in niche markets, localized to the places of agricultural production.

But like green chemistry, this biocapitalism calls into question the organization of research and development established during the twentieth century with large petrochemical industrial conglomerates in charge of innovation. It is giving way to another model in the form of small biotech start-ups – some of which are listed on the stock exchange – that carry out upstream research using private and public funds in order to secure licences. The licences are then bought by the large industrial pharmaceutical groups, which take care of development, scaling-up, clinical trials and marketing.

More than the organisms themselves – invited to be gently manipulated – the carbon skeleton of the long anti-parallel chains that form deoxyribonucleic acid (DNA) and ribonucleic acid (RNA) is now the source of value. Value springs from the molecular arrangements of polynucleotides with their four bases – adenine (A), cytosine (C), guanine (G) or thymine (T) – as amino acids that form proteins. The famous DNA double helix, considered in the middle of the twentieth century to be the universal basis for heredity, has become an array of machine parts. What Francis Crick liked to call 'the secret of life', the explanation of life in physical-chemical terms, is now a tool that can be synthesized in the laboratory. Molecular recognition to match the sequences of the two strands has been used for some decades now for carrying out nano-transistor self-assembly. In the early twenty-first century, recombinant DNA became the main starting point for a variety of mutation techniques used to build living machines. Carbonaceous macromolecules, often compared to Lego-type blocks, are treated as modular molecular bricks that can be arranged or rearranged to shape living things 'by design', tailored on purpose. With building blocks that can be standardized and made interchangeable, we are starting to fabricate life. Everything seems possible again. While the entire industrial and economic world recognizes that stocks of fossil fuels are limited, the combinatorial resources of living organisms are raising hopes for unlimited growth based on renewable resources in a finite world.

For example, metagenomic analyses of the communities of unknown micro-organisms that inhabit most biotopes are now being carried out.[13] The genes present in a solution are identified and annotated by assigning them a function by analogy with another sequence already annotated in genomic databases. But the main objective is to create living machines capable of performing a specific task, for example degrading cellulose. The gene sequences responsible for this functionality are then sought out and inserted into a yeast genome using targeted mutagenesis techniques. Computer simulations of these flows and analytical methods – mass spectrometry, nuclear magnetic resonance (NMR) spectroscopy and liquid chromatography – enable one to predict and maintain performance. This is the challenge of synthetic biology, a fast-growing discipline that aims to re-design or re-engineer living organisms, with tantalizing innovations on the horizon. There is the promise of biofuels made by microalgae or yeast, bacteria that produce medicines or clean up pollution. One of this new discipline's objectives is to build artificial micro-organisms from a minimal cell treated as a chassis to which functionalities are connected. In all instances, the functioning of these cellular factories presumes the control and optimization of the metabolic flows of carbon that lead to the desired product.

A universal machine

In this new bioeconomy, the carbon chains of nucleic acids are seen less as material structures than as sequences of information that code or send instructions for making amino acids and orchestrating the ballet of enzymes and proteins inside the cell. DNA is a code, a program that can be written, read and rewritten in a few letters thanks to high-throughput sequencing platforms, in which we can cut and paste with 'molecular scissors' to re-edit sequences as we would edit a text. This technique, known as CRISPR-Cas9,[14] comes from hijacking a bacterial immune tactic. Bacteria integrate into their chromosomes rewritten fragments – palindromes – of sequences from phages and viruses that have infected them, and synthesize proteins (Cas9) capable of recognizing and cleaving these sequences if they are reinserted into their genome. Taken from the evolution of micro-organisms for which it constitutes a sort of epigenetic memory, this immune tactic is transformed, on the bench, into a targeted genome rewriting technique. Thanks to the various techniques of genome editing and targeted mutagenesis, bacteria can be transformed to produce molecules of interest to us. For example, at the cost of some forty gene mutations requiring the collaboration of the DuPont group and a start-up company, Genencor, the *Escherichia coli* bacterium was persuaded to produce 1,3-propanediol (PDO).

Organisms reprogrammed for technical or medical purposes are not the product of history or evolution; they come into existence as the result of a design process, qualified as *rational design.* Like the synthetic polymers produced by petrochemistry, these carbon compositions escape natural time and are only written into laboratory space-time. They cannot survive in an open environment, with all the biodiversity of natural environments. Their isolation is an essential condition for the success of these biotechnologies. Every effort is made to confine the modified micro-organisms in an enclosure, to prevent them from spreading in the natural environment and from competing with natural micro-organisms. In short, to keep them out of the world.

The ambition of certain champions of synthetic biology is to do *better* than evolution, or *differently* from evolution, to create other forms of life; life as it could be. The horizon of possibilities opened up by the multiple combinations of molecular sequences seems more attractive than the world inherited from evolution. George Church of Harvard Medical School, for example, is proposing to resurrect the woolly mammoth from the genome of the Asian elephant in order to repopulate the tundras. This research programme is one step in a vast project outlined in a book entitled *Regenesis*, which consists of re-creating living things from the dawn of the first living things to the manufacture of posthumans who will be invulnerable to all viruses.[15] Such a project implies treating DNA as a machine that can build all possible or imaginable machines, a universal machine:

> Just as computers were universal machines in the sense that given the appropriate programming they could simulate the activities of any other machine, so biological organisms approached the condition of being universal constructors in the sense that with appropriate changes to their genetic programming, they could be made to produce practically any imaginable artefact. A living organism, after all, was a ready-made, prefabricated production system that, like a computer, was governed by a program, its genome. Synthetic biology and synthetic genomics, the large-scale remaking of a genome, were attempts to capitalize on the facts that biological organisms are programmable manufacturing systems, and that by making small changes in their genetic software a bioengineer can effect big changes in their output.[16]

Thanks to this universal machine, synthetic biologists can play God. And they do not hesitate to do so in order to create a sensation or attract public attention. However, it is less about playing God than about playing with the possibilities offered by living molecules and sensationally breaking records. In this game of exploring the possible, synthetic biologists once tried to 'correct God's mistakes'. More specifically, Steve Benner, one day at the University of Gainesville in Florida, was surprised that electropositive charges followed one another in the carbon skeleton of DNA. He pointed out the gap between the products of contingent evolution and the products of *rational design*. In a lecture delivered at The Pittcon Program 2012 Conference, Capstone, boldly entitled 'Redesign DNA: Fixing God's mistakes', Benner argued that the structure of DNA is far from perfect because it reflects events that occurred at the origins of life and in later episodes of the earth where the constraints on the structure of biomolecules were quite different from what they are today, and certainly not prepared to create a biomolecule that serves the purposes of analytical chemists, bioengineers or synthetic biologists. So, Benner and his team tried to replace riboses with more flexible glycerols. They put all their art into synthesizing a more perfect molecular system, one that was more rational, more respectful of the elementary laws of electrochemistry with alternating positive and negative charges. And what do you think happened? The molecule collapsed. It did not hold and could not support evolution.[17] This is a failure full of lessons: it reveals the essential role of sugars in the formation of the double helix, or, as Benner jokingly points out, it indicates what is really a necessary condition for biological evolution: a carbon skeleton that doesn't give a damn about electrochemistry.

It is clear that none of the forms of sustainable economy envisaged to date can do without carbon. Whether we limit ourselves to drawing on the carbon structures offered by plants – treating plants as chemical reactors – or whether we exploit the capacity of the building 'bricks' of life, self-assembling and organizing matter in white biotechnologies, we are always banking on the potential of carbon to form robust structures capable of evolution.

14

The carbon market

When deliberating in assemblies and tribunals, the Romans used to record their opinions on tablets. They used white, with chalk, for approval or acquittal, and red, the colour of embers, to express disagreement, blame or infamy. Hence the French idiom *marquer au charbon* ('marking with coal'), red embers being the visual metaphor for burning coal – in Latin: *carbone notare*.[1] Today, we 'mark with carbon' our climate misdeeds by leaving a black mark on the environment, our 'carbon footprint'. Expressed in tonnes of CO_2 emitted per activity, or hectares of forest required to absorb these emissions and transform them into organic carbon, the imperative is to mitigate this black mark. The call for sobriety and the development of 'carbon-neutral' services is accompanied by 'carbon audits', which have led to the development of a system of compensation, or more literally market-based 'carbon offsetting', where emission credits are exchanged for the right to continue to emit. Since the Kyoto Protocol, ratified in 1997 by eighty-four countries, greenhouse gas emissions have been assigned a price, which is variable, but established on the basis of carbon taken as the unit of account and exchange. So carbon does not just circulate in the air, in living organisms and in fire-powered machinery. It also flows through expert groups, international deliberative assemblies (Conference of the Parties or COP), and finally through financial transactions that are trying to deal with global warming.

This chapter explains how carbon is becoming the focus of a vast accounting and regulatory scheme for exchanges between humans and the biosphere through the carbon market. How did the problem of global warming become an economic and financial issue? How can a chemical element become an international currency, a creature of the market, and what are the upshots of this? Is financial capitalism capable of cooling the planet? And are these two modes of carbon existence – ecological and financial – soluble in each other?

Carbon finance

In 2015, a report by the International Carbon Action Partnership (ICAP) counted nearly twenty carbon markets, representing 40% of the world's gross domestic product.[2] These markets are of two types: institutional or voluntary.

These institutionalized markets, such as the Regional Greenhouse Gas Initiative in the north-eastern United States or the European Union Emission Trading Scheme (EU ETS) in Europe (the largest carbon market worldwide), are regulated by international organizations, in particular the United Nations.[3] The main sectors concerned are the companies producing energy and 'heavy' materials such as cement and steel. Since the regulated market was set up in 2005, each signatory state of the Kyoto Protocol has allocated (free of charge or by auction) emission allowances to its highest greenhouse gas emitting companies; companies that exceed their allowances can either buy emission rights from those that have capitalized on them or acquire 'carbon credits' with which they fund projects to reduce greenhouse gas emissions or absorb CO_2 in developing countries.

This market is based on the observation that reducing greenhouse gas emissions has a cost for manufacturers in terms of investment and/or loss of earnings.[4] As long as the price of emissions is higher than the cost of reductions, reducing is equivalent to making a profit, brought about through the acquisition of emission rights, which are sold by the most 'virtuous' companies to those that have exceeded their ceiling. After that, the price of carbon evolves according to supply and demand.

The voluntary carbon trading market operates without allocating quotas, on the basis of standards and quality labels governed by an ISO standard (No. 14064). It is run by holding companies such as the Chicago Climate Exchange, private exchanges such as the now-defunct BlueNext or the Carbon Trade Exchange and foundations such as CarbonFund or MyClimate. Companies participate that are not legally subject to the institutional market, such as transportation companies, or are from countries that are not signatories to the Kyoto Protocol, such as the United States. However, *anyone* can buy carbon credits on a voluntary basis to help finance an energy efficiency, renewable energy or tree planting project in a developing country.[5] Considered a donation, the voluntary purchase of carbon credits is generally tax deductible.

Carbon markets have given rise to a host of new players in green finance: offset providers, carbon credit brokers and traders, carbon traders, carbon cowboys (a pejorative term for rapacious entrepreneurs who launch low-quality offset offers to make a quick profit); practices such as carbon pricing, carbon branding (labelling a product or event 'carbon neutral' through offsets) or carbon banking (pocketing the proceeds from the sale of surplus emission permits).

Even if the barrier between the regulatory and voluntary markets is relatively watertight, some carbon trading platforms offer mixed packages or are partially integrated into the institutional market, for the auctioning of quotas, for example. As for the institutional market, it may be 'regulated', but it is no stranger to the usual deregulation practices in finance, such as forward buying and selling, short selling, derivatives or 'structured' products (whose value depends on the evolution of shares), securitization and 'hot air' sales (of credits already used on the market). It attracts financial rogues and lends itself to scandals, such as the gigantic carousel fraud (companies set up in different countries carry out fictitious resale operations at a loss with reimbursements of overpaid VAT[6]). Some of the biggest buyers of carbon credits are banks such as Barclays, Goldman Sachs and Credit Suisse.

The new general equivalent

Carbon is causing a buzz, where millions of bees – economists, financiers, climate scientists, politicians – are busy slicing, calculating, reporting, valuing, scripting, monitoring, monetizing and buying up activities that contribute to global warming. This myriad of operations is dependent on the establishment of a single general equivalent, the 'carbon equivalent'.

The term 'carbon equivalent' can be understood in several senses. First, it is a *unit of measurement* that makes it possible to compare a heterogeneous population of substances, the greenhouse gases; it is also a *currency for the exchange* of accounting tokens – credits and permits – making it possible to create securities, debts and transactions on an emissions market. Finally, and above all, it is a *postulate*, in the sense that, far from describing equivalent phenomena, it *establishes* an equivalence among them, making it possible to take action to bring them back into balance. A generalized equivalence postulate.

The carbon equivalent makes it possible to compare gases as different as, for example, nitrous oxide (N_2O) and sulphur hexafluoride (SF_6) by considering them in terms of their contribution to the greenhouse effect. Their climate warming potential is calculated relative to CO_2 as the 'reference gas'.

Once these gases are made comparable with each other through the carbon equivalent, they are in turn equated with *human actions*. Actions as diverse as planting a tree, distilling oil, grilling a steak, eating organic food, building a wind farm, switching energy providers or travelling by train or plane are all part of this equivalence system. Activities that take place at distant points of the globe or in the calendar are all included on the same balance sheet thanks to this carbon currency. The only thing that matters is *global* reduction for the climate, not where the reductions are made.

Thanks to the carbon equivalent, not only gases and actions are taken into account, but also commitments, or promises of reduction, sealed by a transaction and considered as 'avoided' emissions, elsewhere and in the future. Hence the decisive importance of discount rates, as in the stock market.[7] Postulated here is the equivalence between *emitting* and *avoiding emissions*. It is equivalent to emitting ten tonnes of CO_2 and buying ten carbon credits (1 credit = 1 tonne of CO_2) corresponding to the quantity of greenhouse gases that *could have* been released into the atmosphere.

This equivalence between actual and avoided emissions is the basis of the flagship measure, 'carbon offsetting', which consists in offsetting incompressible emissions by financing carbon projects designed to reduce or avoid other *equivalent* emissions in a distant place and time. Thus, an emitting activity (car or air travel, domestic heating, industrial production, etc.) is declared 'carbon neutral' when it is *offset* by an *equivalent* investment in some reduction project (forestry, wind farm, etc.).

Within the institutional market, carbon offset trading introduces 'flexibility mechanisms', which allow developing countries such as China and India to be exempted from caps while allowing capped countries to offset their emissions to meet their Kyoto targets. According to the principle of equivalence, it does not matter where the reductions are made, as the result is considered on the scale of the global climate without taking into account the territorial inequalities that global warming brings about (disparities in housing types, urban heat islands, rising water levels or water shortages, climate migrations, new statelessness[8]).

A common measure

How can carbon make such global balances possible, aggregating elements as diverse as chemical substances, daily activities, ecosystems and factories?

For each of the greenhouse gases, a 'global warming potential' (GWP) is estimated. The GWP is an index for comparing the contribution of a gas to global warming *relative to that of CO_2*, whose GWP is conventionally set at one over a hundred years. Carbon dioxide is therefore used as the standard for determining the GWP of other gases. It acts as a commensurator: it provides a common measure. For example, the GWP of nitrous oxide (N_2O) has been recently estimated to be 273 over a hundred years (a value periodically revised). It means that it has a warming potential 273 times greater than CO_2 and that a company emitting one tonne of N_2O will be counted as emitting 273 'tonnes of CO_2 equivalent' (tCO2-eq). To convert the volume of emissions of a gas into tCO2-eq, simply multiply the n tonnes emitted by its GWP.

It remains to be seen how the GWP of a gas is determined. It is not something one would measure in the air. The GWP is a *hybrid index*, incorporating scientific *and* political data. On the one hand, it is based on studies of the influence of a particular gas on the 'radiative balance': the calculations of radiation absorbed/remitted by the earth/atmosphere system at the boundary between the troposphere and the stratosphere, expressed in watts per square metre. When a gas contributes to upsetting this balance, it is referred to as 'radiative forcing'. Positive radiative forcing (more energy received than emitted from the system) leads to global warming. The measurement of radiative forcing, which is already relatively complex because it depends on the state-of-the-art of climatological knowledge about forcing factors and prognoses about the chemical fate of gases in the atmosphere, is then integrated into the formula for calculating GWP:

$$GWP(x) = \frac{\int_0^{TP} a_x[x(t)]\,dt}{\int_0^{TP} a_r[r(t)]\,dt}$$

The formula balances the gas to be assessed, x, against the reference gas r, i.e. CO_2. At the top of the fraction bar, a_x is the radiative forcing of 1 kg of the gas under consideration; $x(t)$ the mass of gas remaining in the atmosphere at time t from an emission of 1 kg of that gas at time zero; and TP the time period chosen to perform this integral. At the bottom of the fraction bar, the denominator shows the same integral for CO_2, noted r for 'reference'. The GWP therefore means 'how much more (or less) radiative forcing a gas is doing than an equal amount of CO_2 emitted at the same time over the given period'.

This balance is established on the basis of spectroscopic measurements accumulated over decades in databases, but also on the basis of models built from given scenarios, which themselves depend on possible and desirable action plans (which may include health and ecological issues other than climate change, as is the case for black carbon or chlorofluorocarbons – CFCs). Not only can the GWP vary according to the period considered, but the atmospheric lifetime itself depends on the initial conditions, which are not taken into account in the calculation of the GWP; they are based on a hypothetical time zero, *ceteris paribus*, 'all other conditions being equal'. The GWP is therefore an approximate index, but one that is perfectly suited to its function; it allows for comparisons in order to guide action. It is nothing more than the mathematical translation of a scenario, a means of bringing together those involved in the fight against climate change on priorities for action.

As sociologist Donald MacKenzie points out, GWP and greenhouse gas 'exchange rates' are designed to function as 'black boxes'.[9] They need to be able to be manipulated by carbon traders with no particular scientific expertise. This 'black boxing' is necessary for two reasons. By avoiding constant renegotiation of conversion rates – only the price per tonne of carbon equivalent fluctuates – it gives the carbon market the liquidity it needs to function and protects it from political or economic disruptions.

Why carbon?

After water vapour, CO_2 is the gas that contributes most to the green-house effect in terms of emission volumes: 39% compared with 55% for water vapour, whose atmospheric lifetime is, however, very short, the time of water being the time of weather. But CO_2 is far from being the most virulent in terms of its global warming potential; it is largely outranked by other carbons. Take methane, CH_4: its GWP is about thirty times greater than carbon dioxide over a hundred-year horizon, and about fifty times greater over twenty years. Black carbon surpasses it by far in the short term, with a GWP over three thousand times that of carbon dioxide over twenty years and eight hundred times over a hundred years. Halocarbons are more virulent over the long term. Among them, carbon tetrafluoride, CF_4, the most widely used gas for plasma etching in the microelectronics industry, has a GWP up to seven thousand times greater over a hundred years. But carbon dioxide is just as outranked by non-carbon gases. Nitrous oxide, NO_2, as we saw, has a GWP of 273 over a hundred years. Sulphur hexafluoride, SF_6, beats all records, with a GWP of twenty-two thousand times greater than carbon dioxide over a hundred years. So why CO_2? Why was it chosen? In part because of the role of a converter between nature and society that carbon has played all through the common history we share with it.

Apart from providing the lowest common denominator (which is convenient when choosing a reference standard), CO_2 is above all a common denominator of human activities – extraction, manufacturing, food production, transport, heating, lighting – all of which can be associated with an emission of this gas. Secondly, carbon dioxide is, as we have seen, the historical archetype of all gases, 'the' gas, *par excellence*.[10] Moreover, the 'carbon footprint' carries the idea of a *trace* left by human activities on the environment. The carbon footprint metaphor, which is so familiar today, suggests desecration (prints left by treading on virgin soil and disturbing natural balances) and at the same time naturalizes the human footprint as the trace imprinted *in the soil*, inscribed in nature's processes, which is itself represented as a carbon storage or reprocessing plant. The carbon equivalent allows human activities to be measured on the scales of natural forces, and thus allows the invention of the concept of the Anthropocene.[11] Finally, because carbon dioxide results from combustion, it emphasizes the choice of fire as the original sin on which our civilization was built, with the resulting 'carbon footprint' as a stain to be erased. Therefore, carbon dioxide connects more than any other gas the issue of global warming with that of our fatal dependence on fossil fuels, our industrial history with the earth's history. In short, CO_2 connects scales of space and time which, without it, would remain distant and incommensurable with our activities.

The fact remains that CO_2 is only one of the greenhouse gases, 'one among many'.[12] This is why the CO_2 equivalent is in turn convertible into a 'carbon equivalent', obtained by multiplying the CO_2-eq of a gas by the mass of carbon atoms contained in one kilogram of CO_2: 0.2727. If one kilogram of CO_2 is worth 0.2727 carbon equivalent, one kilogram of methane is worth 6.27 carbon equivalent, nitrous oxide 81.27 and sulphur hexafluoride 6518.2. It is then the carbon element that gives us the 'exchange rate' of greenhouse gases.

The choice of the carbon element rather than CO_2 as the ultimate equivalent has a twin paradoxical effect: towards the concrete and towards the abstract. On the one hand, the carbon equivalent aims to bring the economy 'back to earth', standing on its feet, to reintroduce it into the material ecosystem that conditions it and that it affects in return. It thus reminds us that what is exchanged and circulating in biological, biochemical, geochemical, geophysical and atmospheric processes is the element carbon.[13] It therefore seems legitimate to consider it as the currency of exchange between human industry and planetary processes.

On the other hand, this second convention introduces the abstraction inherent in the institution of a currency of exchange, because a class of material objects (shells, wheat, metal, paper, etc.) takes on a universal value by making its materiality disappear in the form of exchange. Its role becomes comparable to that of the gold standard in the past. Carbon then takes on the qualities of 'material abstraction' or 'symbolic materiality' that characterize money or currency.[14] Instead of bringing the economy down to earth, it tends to make ecology a speculative product like any other.

But are carbon flows, now the material currency of our trade with the biosphere, soluble in a dematerialized and financialized economy? And at what price?

The price of carbon

'Offset credits are an imaginary commodity based on subtracting what you hope will happen from what you claim would have happened,' says a sceptical commentator.[15] As Augustin Fragnière, a researcher in environmental philosophy, points out, the very term 'offset' improperly suggests that a perfect balance is reached between emitted and avoided gases.[16] However, we have seen that offset mechanisms assume counterfactual scenarios of expected 'saved' emissions that are necessarily uncertain – when they are not deliberately inflated by the players. Offsetting is based on the idea of a balance to be reached between two antagonistic elements in terms of values. Its psychological attraction lies in a kind of moral balance that exploits the 'cognitive dissonance' effect between the concern to preserve the climate and the awareness of contributing to the greenhouse effect through one's activities. To reduce this dissonance, actors have the choice between modifying their behaviour, acting on their environment or cancelling the dissonance. The last option is the solution proposed by offsetting. It allows an actor to do something about the planet by delegation, without taking on the consequences and requirements of climate action. It neutralizes the dissonance by restoring the lost compatibility between behaviour and concern for the climate. It is the least costly solution in cognitive terms, since its cost is reduced to its financial cost, which is often presented as 'competitive' and therefore ridiculously low.

According to Fragnière, it should be stated that, strictly speaking, emissions and reductions are *comparable*, not equivalent. Offset markets instrumentalize carbon by turning into a black box its ability to provide a common measure. As a commensuration device, carbon allows for comparison; it offers a diversity of possible standards whose determination should be open to collective deliberation, in the same way that there are local or alternative currencies and debates on the relevant indicators of wealth. The postulate of global equivalence ignores the role of carbon as a means of comparison and nips in the bud any discussion of how carbon should be handled. While carbon offers itself as a mediator of exchanges between human activities and natural processes, offset markets are fuelled by contempt for carbon and make it the enemy to be reduced to nothing, to be 'neutralized' or 'sequestered'. Carbon has gone from being a common measure to a common enemy.

Another major difficulty with the carbon market is that it ignores the contradiction between reversibility of benefits and irreversibility of damages. While a volume of greenhouse gases 'waits' to be offset – for example, while the trees grow – it continues to contribute to global warming and cause irreversible effects. Once the trees have grown, the emission is offset, the emitting activity declared 'carbon neutral'. However, the result is not the same as if these emissions had never existed, which is what the term 'offset' suggests.

The term 'carbon neutral' is even worse, as it suggests impunity. Any carbon-neutral person would have nothing to be ashamed of. It is a climate absolution device; carbon offsetting is often compared to the system of indulgences, the forgiveness market set up long ago by the Catholic Church.[17] Cashed-up climate sinners can wash away their sins (*greenwashing*) by financing 'green deeds' carried out elsewhere and at other times, i.e. emission reductions that they themselves do not have to work to get.

The carbon market aims to re-synchronize the carbon tempos mentioned in the previous chapter. It plays on an image of communicating vessels, as in a lab, syphoning up the harmful effects of industrialization by setting up a monetized and hyper-fluid exchange system for beneficial action. The carbon equivalent functions as a postponement tool, for putting things off to another time and place. This distancing in time and space of emission reductions leads to a loss of contact with the concrete aim of combating global warming, i.e. the effective limitation of massive carbon oxidation and its re-metabolization in the form of organic carbon. It can lead to a feeling of unreality of the climate threat, especially when the act of offsetting is done with a few clicks on the internet in a well air-conditioned office. Finally, this postponement of action in a market of promises poses the problem of intergenerational justice.[18] Boasting about being 'carbon neutral' when the concrete reduction measures are still waiting to be put into effect is tantamount to clearing oneself of a debt to future generations, but it actually implies an extra burden of action in the future, since the next generation will have to take responsibility for its own emission reduction actions in addition to ours, which we will have postponed through offsetting. The carbon market therefore poses a problem of relation to time, or more precisely a problem of relation between multiple times.

While claiming to reconcile the multiple and conflicting temporalities of carbon – the linear time of growth, the looping time of carbon cycles – it brings them all down to the present.

The tradable permit system went through long periods of instability, if not failure. In order to begin to influence industrial investment, the price of carbon would have to remain at a certain level. However, the system is working in the wrong direction, since the market sets the price. The first phase of the EU's ETS (2005–7) was a monumental failure, as it saw the price per tonne of CO_2 fall from €25 to less than €1 due to an initial over-allocation of free emission allowances that encouraged short-selling and downward speculation. During Phase II (2008–12), the banking and financial crisis caused the carbon price to plummet, despite a slight recovery due to the Fukushima accident. Since 2012, the price of a tonne of CO_2 has fluctuated between €5 and €7, which is far too low to encourage investment in emission reductions.

Following the collapse of the carbon market, hope has been pinned on moves to set a *global reference price* for carbon by a coalition of seventy-four countries and over one thousand companies (Carbon Pricing Leadership Coalition) officially launched on 30 November 2015 at the opening of COP21 in Paris. The reference price for carbon would allow the various players – industry, finance and government – to agree on a 'clear economic signal'. It is the subject of skilful calculations that involve three carbon identities: (1) the socio-economic price must anticipate the cost of all the damage caused by the emission of a tonne of CO_2 and is based on the trope that it is carbon that is guilty of damage to the economy and society; (2) the abatement (or avoidance) price is the price to be paid to reduce a tonne of carbon emissions and is based on the trope of carbon as a nuisance to be avoided; (3) and the effective carbon price is based on an assessment made by policy makers or voluntary behaviour of public and private operators at a given time in a given sector.

Determining the socio-economic price is particularly delicate. Firstly, equivalence coefficients must be defined for each of the greenhouse gases, while factoring in the fact that their radiative intensities and lifetimes in the atmosphere are extremely variable. Secondly, it is necessary to assess all the consequences of the damage caused by carbon emissions: reduced growth, migration, etc., which is a controversial issue. The price is therefore the result of a compromise between the opinions of various stakeholders on the political role to be given to future generations.

Setting the abatement price also implies anticipating the future cost of alternative technologies to limit CO_2 emissions. However, this price varies considerably depending on the choices made at a given time. Hence the highly contrasting scenarios proposed by the IPCC's Group III on climate change mitigation. Moreover, the socio-economic price and the abatement price imply comparing costs or revenues at different periods in time, which may be quite distant. The choice of the discount rate is therefore decisive.

Only the effective price can be calculated directly. Theoretically, in an ideal situation with optimal regulation, it should be equal to the socio-economic price. But this is far from being the case, which is why the difference between the two is used to evaluate the effectiveness of public policies.

In April 2017, the French Académie des technologies recommended a reference price of €50/tCO2-eq, to be periodically adjusted, and subject to public debate, to ensure equal treatment of future generations.[19] But to achieve the COP21 objectives, other mechanisms need to be added to the setting of a reference price, such as subsidies to encourage energy saving and stricter regulations, like a ban on coal-fired power plants.

Finally, in the opinion of the majority of states and private sector players taking part in these negotiations, translating this reference price into imposing a *carbon tax* at the global level would be anti-liberal, and thus unworkable.[20] The idea in vogue is therefore to create a *carbon debt* that converts the environmental debt of states into financial terms, possibly accompanied by a bonus/malus system. Some economists even recommend merging all carbon markets under the aegis of this debt.[21] Where carbon taxes and emission permits require immediate payment of the socio-economic cost, the 'carbon debt' mechanism relies on progressive repayment, 'like paying off the mortgage on a house rather than paying cash'.[22] Mortgaging the biosphere, this would be the essential proposal from climate financiers.

However, these speculations have come to an end – perhaps temporarily – since the price of carbon rose again in 2018 to finally exceed, for the first time in its history, €100 in spring 2023. This is a price that is not abnormally high, but in line with the price expected if the market were efficient – which remains questionable.

This recent rise is due to many factors. The most optimistic advocates of the carbon market as a form of 'sustainable finance' attribute this recovery to the 'rational anticipation' of economic actors following the COP21 Paris Agreement on climate change (2015), the adoption of the European Green Deal (2020) and of its subsequent European Climate Law (2021), setting a legally *binding target of net zero greenhouse gas emissions by* 2050. Accordingly, these 'green' market actors would be not only rational but also reasonable; they would keep the price of carbon high in anticipation, in their great wisdom, of the announced end of the reign of carbon.

A more realistic factor that played out is a structural reform making the trading system more restrictive, namely the introduction of a Market Stability Reserve (MSR), set up by the European Commission in 2019 to remedy the problem of excess allowances in circulation. Before that, the economic actors were often 'black-loading' their quotas (a practice whereby allowances are temporarily withdrawn from auctions and then re-injected) or doing 'setaway' (a similar trick but with allowances re-injected at an unspecified date . . .). Yet these practices induced a surplus of supply, which drove down the price of carbon. The MSR temporarily withdraws the surplus of allowances in circulation and re-injects them into the market based on definite indicators.

Other factors are more contextual. The recent post-Covid economic recovery has mechanically increased the volume of emissions and consequently the demand for allowances. In addition, the rising prices of oil and then of gas due to the Russian invasion of Ukraine have favoured a strong comeback of the production of electricity from coal, and have consequently led to an increase in demand for emissions allowances. As we can see, the price of carbon does not just act as a thermostat, and the recovered health of this 'price signal' is not necessarily a sign of good climatic health.

The carbon market is often denounced as a 'techno-fix'.[23] This term refers to any measure that offers technical solutions to environmental, ethical or socio-political problems, often induced by older technologies.[24] The 'fix' metaphor evokes patching up a worn-out or wonky machine, but also injecting a narcotic. Our current lifestyles are so addictive that we shoot up on innovation in order to maintain them at all costs. We should be tackling the causes, changing our behaviour, our lifestyles, production and consumption patterns. But we prefer to 'fix' the detrimental consequences of old innovations with yet new ones, whose negative consequences never outweigh the prospect of capitalism running out of steam in the risk–benefit balance. Geoengineering, when it proposes to change the chemical composition of the oceans or to put a space parasol into orbit to cool the planet, is often cited as an extreme example of a techno-fix. But carbon offsetting is also a techno-fix because it allows a technical measure (offsetting the effects of emissions) to be substituted for a change in behaviour (reducing emissions).

However, the opposition between technical and behavioural solutions cannot be the last word in the criticism of the carbon market. Not only are these two attitudes not opposed in themselves, but practical ecology actually requires a more developed technical culture in everyday life. Building a wind turbine in your garden, recharging your batteries by pedalling, composting your organic waste or practising permaculture is as much about behaviour as it is about techniques, and requires skills, instruments and a constant effort to 'repair' the world and its living things. Technologies and life deserve a *joint culture* rather than a common market. Yet such 'green' finance tools maintain and even increase the gap between technologies and lifestyles. They do so by choosing market-fix rather than techno-fix, by subjecting climate action to a division of labour between intentions and actions. Those who claim to be 'carbon neutral' are not those who are actually carbon neutral. The alternative, then, is not between behavioural change and technical change. Rather, it is about resisting their dissolution into a third kind of change that is not really a change at all, the perpetual financial momentum; it involves activism towards an emergent technical carbon culture in which lifestyle changes and technical changes would be intermeshed. But to nourish this association, a third ingredient is required: carbon narratives.

This means taking advantage of the writing devices that this nature–culture converter offers humans as ways to inscribe their history in scales of space and time that go beyond, and underneath, the human.

PART III

CARBON TEMPORALITIES

15

Carbon cosmogony

The first part of this book told the story about the invention of carbon in its multiple chemical modes of existence. The second part showed how, thanks to carbon, human art, culture and industry developed and it discussed the kinds of technical or economic cultures that gathered around the element. This third and final part is devoted to the cosmic, geological and biological modes of existence of carbon. It addresses the issue of climate change. It outlines a critique of the concept of the Anthropocene, which places 'Man' at the centre of a single, linear and universal time perspective, whereas the various modes of existence of carbon invite us to take multiple temporalities into account and to think of the earth system as polychronic.

Let us transport ourselves to a distant time, well before the time of humans, even before the existence of the planet earth, the solar system and our galaxy, and, from the genesis of the cosmos, follow the steps carbon has taken. Far from following a linear path through scales of space and time neatly nested along a single axis, we will discover a multiplicity of images of time, all contingent. We will see how the world, in the sense of the cosmos first of all, then of the earth, as a veritable ball of interwoven time, with its cycles, climates and beings, is inventing itself with carbon.

In the mists of time

So here, then, is a new biography of carbon, a scientific and speculative account of its origins that gives it a date of birth. Not from the 'Big Bang', the singularity that marked the birth of the currently observable universe 13.8 billion years ago,[1] but when the first massive stars formed about a hundred million years later. In the few femtoseconds following the Big Bang, elementary particles are generated; in a microsecond they form molten protons in a hot plasma that also contains photons and electrons in constant interaction. As the universe expands at high speed, it cools down; matter and radiation are being decoupled; below 3,000 Kelvins, protons and electrons combine one by one and form stable, electrically neutral hydrogen atoms. Photons are released; the universe begins to become visible. The first elements are generated in the first few minutes after the singularity. Hydrogen, consisting of a single proton with a single electron orbiting it, makes up three-quarters of the mass of the universe, helium one-quarter.[2] The rest of the primordial universe contains minute traces of elements such as deuterium,[3] lithium and beryllium, but no heavier elements. Fifteen minutes after the Big Bang, primordial nucleosynthesis is extinguished. The temperature of the universe suddenly drops and its chemical composition is stabilized.

Time passes. A million, two million years. Or rather it does not pass – it depends – because not much happens. The universe, crossed by faint glows and a few X-rays, expands slowly. Gravitational waves echo the initial event and timidly shake up the texture of space-time. Events are rare: a few collisions of atoms, a few spin inversions of the hydrogen electron[4] emitting a photon. The universe is cold, traversed by clouds of wandering atoms. Its electromagnetic activity is weak.

Around a hundred million years ago, certain nebulae of atomic gas acquire sufficient density to form one of the rare molecules of which the primordial universe is capable, dihydrogen (H_2). From these molecules, the first stars are born, from rare cosmic events whose probability increases with time: gravitational collapse of the clouds on themselves, movement of a shock wave through another cloud of hydrogen, formation of a first star, then this star moves close to another cloud of gas ... The contraction of the hydrogen atoms raises the temperature until their nuclei fuse; the energy released by the nuclear fusion counteracts the gravitational forces until a first equilibrium is reached: a star is born.

During the first billion years of existence of the present universe, the first stars, then the first galaxies and black holes emit X-rays and ultra-violets that heat up and excite the primordial hydrogen atoms by tearing off their electrons. This leads to the ionization of interstellar hydrogen gas, which in turn increases the potential for star formation. The cosmos lights up.

The stars are burning hydrogen and converting it into helium. Low-mass stars release their outer layers, which are less affected by fusion reactions; they leave a core of helium that evolves into a white dwarf, a small, very dense, very hot, but dim star. In the most massive stars, including the sun, the temperature of the centre is such that thermo-nuclear fusion reactions of helium are beginning in the outer layers. This is where heavier elements are born of the fusion of helium: carbon, of course, but also neon, nitrogen, oxygen, boron, silicon and even iron, the champion of stability. The process is known as 'quiet' nucleosynthesis. The most massive stars, which are also the hottest, end their lives in a supernova. These supergiants only live for a few hundred million years and eventually collapse in on themselves. A shock wave then sweeps the star from the centre to the periphery and reignites the fusion of the outer layers. This is explosive nucleosynthesis, from which elements heavier than iron are produced. In this way, stars generate elements and distribute them.

If we adopt this grand cosmogonic narrative, carbon should not have existed in the first place. Indeed, when we look for the mechanisms of its appearance, its existence seems quite improbable. Yet carbon does exist, and is a key component in the formation of the universe and other elements. Without it, the standard account of stellar nucleosynthesis no longer holds. Let us examine this in detail.

Unlikely carbon

The thermonuclear conversion of hydrogen into helium takes place in two types of reaction. The proton–proton chain, proposed by Jean Perrin in the 1920s, describes the production of a helium nucleus from the fusion of four hydrogen nuclei. This is the main fusion pathway in stars of mass less than or equal to the sun. In more massive stars, a second reaction takes over as the main source of energy, the CNO (carbon–nitrogen–hydrogen) cycle proposed by Hans Bethe in 1939, in which carbon acts as a catalyst for thermonuclear fusion reactions converting hydrogen into helium, while being regenerated at the end of the cycle (Figure F).

This reaction consists of a main cycle and – depending on the increasing size of the stars – several adventitious cycles coupled to the first one and contributing to the synthesis of oxygen and nitrogen nuclei at the end of a complicated series of reaction loops (Figure G). The CNO

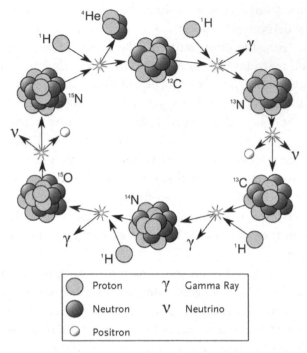

● Proton		γ	Gamma Ray
● Neutron		V	Neutrino
○ Positron			

Figure F. Main loop of the CNO cycle. Carbon-12 (top) acts as a catalyst for the transformation of hydrogen (H) into helium (He) and is regenerated at the end of the cycle.

is thus a cycle that sometimes regenerates carbon by helping to produce helium, and sometimes generates other cycles in which carbon is transmuted into heavier elements – a cycle that makes other cycles, while looping back on itself.[5]

Figure G can be read from the top left corner. The main loop I has a small but non-zero probability of bifurcating into loop II: in which case, ^{15}N no longer gives ^{12}C back by emitting a ^{4}He particle (first arrow going down on the left) but gives ^{16}O by de-excitation of the nucleus and gamma emission (first arrow going down on the right). Thus the CNO cycle can be bicycle or tricycle, sometimes even quadricycle. In all cases, it loops in on itself. The respective importance of these additional cycles depends on the size of the stars.

Carbon is therefore a key player in one of the two major reactions making stars shine. In so doing, it also plays a facilitating role in enabling

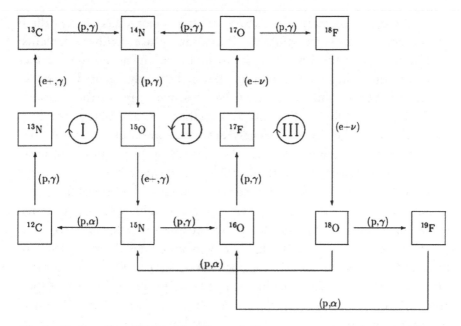

Figure G. Complete CNO cycle. Key: p, hydrogen proton; α, helium-4 nucleus; e +, positron; γ, gamma ray.

stars to produce elements other than itself. But how on earth did it appear there at all?

Historically, Hans Bethe was interested in the problem of energy production in stars and not in nucleosynthesis itself. His theory therefore presupposed the existence of the carbon nucleus, without attempting to explain its genesis. His 1939 paper 'Energy production in stars'[6] does mention a possible mechanism for the formation of carbon-12 via the collision of three helium-4 nuclei (2 neutrons + 2 protons) or 'alpha-particles': 3 × 4 = 12, a reaction that would come to be called 'triple-alpha (α)'. But the probability of a triple collision is *a priori* very low. Bethe then estimated that such a reaction, under the current pressure and temperature conditions of the sun and similar stars, could only yield negligible quantities of carbon-12. The conclusion was that 'there is no way in which nuclei heavier than helium can be produced permanently in the interior of stars under present conditions'.[7] Bethe was happy to set up the problem and leave it as a riddle for others to solve.

Cosmological theories of the 1940s, such as George Gamow's, called this enigma the mass-gap problem. Primordial nucleosynthesis explains well enough how the first elements form by capturing protons in the moments immediately after the Big Bang, but only up to helium-4. Between helium-4 and carbon-12, there is a mass gap. In the periodic table, this gap is filled by lithium (6), beryllium (8) and boron (10), but these light elements are immediately destroyed by fusion reactions and cannot serve as intermediary links. As an essential player in the activity of stars and all known life, carbon should not have existed.

As Bethe noted, the amount of carbon that could be generated by the triple-α reaction is insignificant, *a priori* incomparable with the amounts of carbon-12 currently found distributed in the universe.[8] The triple-α reaction requires sufficiently high temperatures for helium to start to fuse, and only stars of a critical size and mass are capable of doing that. The reaction could take place in red giants, as proposed by Ernst Öpik and then Edwin Salpeter in the 1950s.[9] But this is not enough. Since the probability of a triple collision is very low, the α particles fuse in pairs instead, and the fusion of two helium-4 nuclei gives beryllium-8 (4 neutrons + 4 protons), which in turn fuses with a third α particle to give a stable nucleus composed of 6 neutrons and 6 protons: carbon-12 (Figure H). However, the beryllium-8 nucleus, which is particularly unstable, is destroyed very quickly. Its half-life is of the order of 10^{-16} seconds. It was

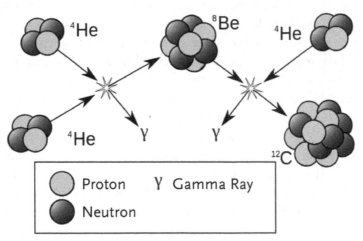

Figure H. Triple-alpha reaction.

therefore highly unlikely that a weak beryllium nucleus would produce a large quantity of carbon-12, according to Salpeter's equations.

After reading Salpeter's article, Fred Hoyle reportedly pointed out to his colleagues that humans are surrounded by carbon in the natural world and are themselves carbon-based life, so stars must have discovered a highly efficient way to produce it, and he was off to find it.[10] He assumed that there must be a previously unknown level of excitation in the carbon-12 nucleus that could resonate with the combined masses of helium-4 and beryllium-8 (according to the mass–energy equivalence of the famous equation $E = mc^2$). If this resonance level existed, the effective cross-section of this reaction would be greatly enlarged and it would become possible to obtain significant amounts of carbon-12 despite the short lifetime of beryllium-8. Hoyle calculated the energy level required for this favourable resonance state to be 7.65 mega-electronvolts (MeV). A few years later, this excited state of carbon-12 was actually produced in the laboratory by William Fowler and his colleagues at Caltech, with a measured value corresponding to the value predicted by Hoyle. This state of the carbon-12 nucleus is therefore called the 'Hoyle state'.[11]

The precise value of this state determines the balance of carbon and oxygen in the universe as we know it (carbon ~ 0.5% / oxygen ~ 1%), and the fine-tuning of this balance in turn determines the possibility of life.

As it happens, Hoyle devised some counterfactual cosmological scenarios.[12] Thus the oxygen nucleus 16 has an accessible energy level dangerously close to the sum of the combined equivalent masses of carbon-12 and helium-4 (12 + 4 = 16). If this resonance level were accessible to carbon-12, some of it would have fused with helium-4 to form oxygen-16. The universe would then have produced much more oxygen at the expense of carbon. Fortunately, the resonance level of the oxygen-16 nucleus is slightly below the Hoyle state (7.65 MeV), at 7.12 MeV. Conversely, if the cosmic productivity of the triple-α reaction had been such that carbon was more abundant than oxygen, a graphite planet would probably have formed near the sun, keeping carbon atoms captive and greatly reducing the chances of the formation of organic molecules, the building blocks of life on earth.

It is clear that without the existence of this 7.65 MeV resonance state of the carbon-12 nucleus, this book would not exist and would have no authors or readers. Thank you, carbon!

Anthropogenic carbon?

Instead of feeling indebted to the carbon that conditions their very existence, eminent physicists have used the existence of this resonance state as a pretext for self-congratulation and for placing humans at the centre of the universe. *If* we are here to observe the universe, *if* our consciousness depends on our biological substrate, *if* our biological substrate depends on carbon, and *if* the existence of carbon depends on the fine-tuning of fundamental physical constants, such as the 7.65 MeV resonance state of the carbon nucleus (but also the force of gravity, the mass of the protons, the charge of the electron, etc.), *then* these constants have to be set from the start to support the possibility of advanced life forms up to the life form capable of understanding the universe, human life.

This long chain of deductions leads to the 'anthropic principle', a term coined by physicist Brandon Carter in 1973 at a symposium on 'the confrontation of cosmological theories with observational data'.[13] In its strong version, it states that the fundamental parameters of the universe are finely tuned *so that* it allows for the birth and development of intelligent observers at a certain stage of its evolution. Since this fine-tuning presupposes an initial anticipation of the final state – humankind – the strong anthropic principle is either a physical version of the intelligent design thesis (this fine-tuning is simply the work of a higher intelligence) or a cosmogonic finalist thesis (the genesis of the universe is governed by an immanent purpose that causes it to strive towards its maximum perfection, i.e. humankind).

Fortunately for the credibility of our eminent physicists, there is a 'weak' version of the anthropic principle that aims to make it scientifically acceptable: i.e. the observable properties of the universe must be compatible with the conditions necessary for our existence as observers, otherwise we would not be there to observe it. It is in this form that anthropists apply Hoyle's prediction. According to them, Hoyle would have supported a weak form of the anthropic argument.[14] Since we exist, the carbon nucleus must resonate at 7.65 MeV (Hoyle's strong version being that this energy level exists *for* us to exist). The experimental success of Hoyle's prediction would thus demonstrate the heuristic power of the principle. It would provide a method of discovery, a guideline for where precisely to look to maximize the chances of observing as yet unknown properties of the universe. Told in this way, this episode is reminiscent of the discovery of the planet Neptune, observed precisely *where* Urbain Le Verrier's calculations had predicted its position to explain the perturbations in the trajectory of Uranus, a typical example of a prediction illustrating the undeniable supremacy of reason over nature. It is only a short step from there to positing humankind as the *reason for the universe.*

In its weak version, the anthropic principle is almost trivial – *almost,* because it implies the plausibility of its strong version and has no other purpose. Otherwise, it is reduced to the statement of a simple abduction procedure requiring the inference of causes consistent with their effects. How, indeed, could one admit fundamental properties of the universe that are not compatible with the properties of the actual universe (which include 'the necessary conditions for our existence as observers')?[15]

Abduction is a mode of reasoning that consists in starting from an observed effect to infer the explanatory cause, as when one infers fire from the observation of smoke. Here, the cause inferred by our anthro-pists is particularly remote because they start from the effect of the 'existence of intelligent observers' (i.e. themselves). But if we are willing to start from the effect of 'existence of carbon', then the prediction of the famous resonance level is simply an inference of the best possible explanation given the assumed initial conditions, as per the principle of abduction. From the observation of smoke, and given the fact that it is hot, that there is wood, that the electrical system is probably old, etc., one infers that fire is most likely the cause of the observed smoke. The only difference is the *a priori* very small probability of the existence of a resonance state of 7.65 MeV and not a tad more or less.

Fine. But probability in relation to what? Relative to the set of possible (i.e. conceivable) universes projected by our intelligent observers to dissipate the sense of anguish inflicted on them by contemplating the radical contingency of their conditions of existence. The existence of carbon, in turn, is a necessary and not sufficient condition of our existence. To bring together all the necessary and sufficient causes that contribute to our existence would be tantamount to summoning up all the chance in the universe – chance being here not the negation of causality but the meeting of independent causal lines. It is this contingency, the result of chance and necessity, that our anthropists cannot tolerate.

Considered as a simple abductive principle, the weak anthropic principle is not really up for debate, except for the fact that it is called an anthropic principle. For why limit such a principle to humans? The observable properties of the universe must be compatible with everything that exists. Since giraffes exist, the fundamental properties of the universe must be compatible with the existence of giraffes; that is certain. But giraffes are not considered to be intelligent observers capable of understanding the meaning of the universe, so there is no talk of a 'giraffic principle', and the same is true of flies, bananas and bacteria. Only the most advanced life form in the universe in terms of intelligence, the physicist working in a suburban laboratory with a super brain and valued colleagues, the cutting edge of humanity and *therefore* of all life forms, can claim this insignificant privilege because 'he' alone is capable of understanding, at last, the meaning of the universe: that it has been made to measure for him.

Discovering that they participate in the cosmos in the strongest sense, i.e. that they depend on its radical contingencies, our 'intelligent observers' demand that their existence justifies that of the cosmos. I exist, therefore the universe is made for me. What arrogance. And how sad. Wouldn't the contemplation of our radical contingency rather be a source of joy? Joy in noting, for example, the improbable links that carbon weaves between the nuclear physics of stars and the organic chemistry of earthlings, of which I am one.

As Hoyle himself said, 'It is not so much that the Universe must be consistent with us as that we must be consistent with the Universe. The anthropic principle has the problem inverted, in my opinion.'[16] In any case, the argument is circular. The universe would be finely tuned so that an intelligent observer (a hypercephalic physicist) could understand that the universe is finely tuned so that an intelligent observer could understand that the universe is finely tuned to . . . etc.

And above all, what a lack of consideration for carbon. A Chinese proverb says that when the wise man points at the moon, the fool looks at his finger. Similarly, our anthropist physicists prefer to see themselves in nature rather than tackle the real mystery: how the carbon nucleus came into existence.

Does the fact that a non-existent element comes into existence thanks to one of its properties – that of being able to resonate at 7.65 MeV – mean that it virtually pre-exists itself? That it is the cause of its own existence even before it exists, the cause of itself? Or does it mean that carbon seizes an opportunity to sustain its own existence? This is the enigma, and it refers to the mystery of the elements and their qualitative heterogeneity. Each element signals a specific relational behaviour that is not completely derivable from the terms of the relation, i.e. the individual physical constituents that pre-exist it. And it is carbon, an exemplary element[17] as a 'relational being',[18] that teaches us this in terms of the history of the universe.

Carbon as earthling

Our planet was formed by accretion with the entire solar system about 4.6 billion years ago. During the initial period of earth's formation, barely a dozen mineral species exist, the most abundant of which is carbon in the form of graphite or diamond.[19] The number of mineral species then slowly evolves during the formation of a still very hot planet, bombarded by meteorites and other materials from the formation of the solar system. The first granites come from partial melts in the mantle. A primitive basaltic crust solidifies. The early history of the earth's rock is one of subduction, melting and uplift, producing new types of minerals. The outgassing of rocky magmas produces an early atmosphere consisting mainly of nitrogen, carbon dioxide, ammonia, methane and water vapour. Soon the earth cools down and forms a crust. Rainfall from the outgassing of molten rock leads to the formation of oceans. At this time, one might count about a thousand types of minerals only. Today's earth has nearly five thousand, a diversity far greater than any other known rocky celestial body. How was this mineral diversity generated?

Through the mediation of life. Life, which requires water, carbon, nitrogen and other elements under specific physical conditions, appeared relatively early on earth. The oldest fossils of micro-organisms detected to date are at least 3.7 billion years old. Today, there are many contesting scenarios for the emergence of life: 'genes first' or 'metabolism first'; 'RNA world' (RNA having replicative and autocatalytic properties) either before or after the partitioning of reactive media into phospholipids forming the basic structure of membranes; the transition from RNA-enzymes to proteoenzymes; the replacement of RNA by DNA as a support for the genome. Whatever the correct scenario(s), all give carbon a crucial, absolutely necessary role. Able to maintain strong carbon–carbon covalent bonds (single, double or triple), it forms long chains with its own atoms (catenation) and constitutes the structural skeleton carrying the other elements of the basic molecules of life (phospholipids, RNA, DNA, proteins). In addition to its structural stability, it has the ability to bind to a very large number of different chemical elements. Its 'sociability' opens the way to an indefinite range of molecular associations, while allowing metabolic dynamics to build on structures that have already been developed and to be sustained over time.

One of the lessons of carbon is that it is not enough to explain how life *arises*; we must also examine how it is *maintained*, how it produces and manages its own conditions of existence. For it is by exploiting the affordances of carbon that life has shaped the planetary conditions that have allowed it, in turn, to flourish.

Thus, the earliest atmosphere contained very little free oxygen in the form of O_2. Left to itself, oxygen reacts with methane (CH_4) to form carbon dioxide (CO_2) and water (H_2O). Under the sea, the first autotrophic[20] micro-organisms oxidize the oceans. They capture carbon dioxide and combine it with water to synthesize the glucose ($C_6H_{12}O_6$) necessary for their metabolism according to the formula:

$$6\ CO_2 + 6\ H_2O + \text{light energy} \rightarrow C_6H_{12}O_6 + 6\ O_2$$

This release of oxygen has a decisive influence on the evolution of the planet.[21] The oceans are quickly saturated with oxygen, and oxygen spreads into the air and becomes one of the components of the earth's atmosphere. Dioxygen is highly exposed to solar radiation, which becomes energetic enough at high altitudes to break down the molecule, a fraction of which converts to O_3, ozone. The ozone layer, protecting the earth and its first living things from ultraviolet radiation, is formed in this way some two billion years ago. But the oxidation of the atmosphere also changes the rocky composition of the earth with the emergence of carbonaceous reefs and ferrous formations. The appearance of eukaryotes, equipped with respiratory organelles (mitochondria) or photosynthetic organelles (chloroplasts) through the symbiotic incorporation of other bacteria (endosymbiosis), and their multiplication via free oxygen, accentuates the transformation of atmospheric carbon into organic carbon. The carbon cycle is set in motion.

At the dawn of the Cambrian period, the land surface is still mostly dry rock. The emergence of land plants about four hundred million years ago alters this surface permanently and leads to the formation of soils. The root system of plants aerates the soil and draws water and nutrients from it for their development. The plant biomass forms litter which returns carbon-rich substances to the soil, feeding bacteria. The Cambrian is an age of an explosion of biodiversity and a period of expansion of mineral diversity with the production and biogenic modulation of all kinds of limestones (composed of calcium carbonates $CaCO_3$ and magnesium carbonate $MgCO_3$) and clays (composed of silicates and aluminosilicates resulting from the alteration of silica rocks such as feldspars). Limestones develop mainly on the sea floor as sedimented accumulation of shells, marine animal skeletons and phytoplankton microalgae. Biomineralization, i.e. the production of minerals by living organisms, plays an important role in terraforming. Living organisms are inspired by minerals and return to them. They produce shells, carapaces, bones and endo- and exoskeletons. It is as if living organisms, building on a molecular scale on a carbon skeleton, are projecting this skeleton outwards into their environment.

Multiple cycles

The amount of carbon, fixed in the earth system, circulates in two cycles: the rapid biological cycle (animals capture oxygen and release CO_2, which is then captured by plants and bacteria that re-oxygenate the atmosphere) is coupled with a geological cycle that takes place on a million-year scale. CO_2 released into the atmosphere by volcanic eruptions or submarine springs is sequestered in the soil as calcium carbonate in sedimentary rocks, which is then dissolved by river and ocean water. The aqueous dissociation of carbonates requires CO_2 and converts it into bicarbonate ions (HCO_3-), while the ions in silicate or carbonate materials are replaced by H+ ions or re-crystallized *in situ*, forming, for example, marble or cipolin in the so-called 'carbonate metamorphic sequence'. The degradation of silicates also consumes large quantities of CO_2, which plays a major role in the evolution of the climate over a long period due to the prominence of these rocks (79% of the earth's rock mass).

These two cycles produce a third cycle, on a median time-scale of a few hundred thousand years, which is called the thermostat effect of CO_2 degradation. As the temperature increases due to the concentration of CO_2 in the atmosphere, the rate of chemical reactions that sequester carbon in the form of sedimentary rocks increases too, thus decreasing the amount of CO_2 in the atmosphere.[22]

The interactions among these cycles are numerous, but their coexistence is not necessarily harmonious. The earth system is not a stable system tending towards thermodynamic equilibrium. It maintains a constant thermodynamic imbalance, and therefore a free energy gradient, metastability and potentials for life. This can be explained not only by the energy flows that pass through it (solar radiation), but also by the activity of the biotic sources that regulate the composition of the atmosphere.

It is therefore not enough to say that life has developed on earth *because* our planet benefits from an exceptional situation, neither too far from nor too close to its solar star.

That would be forgetting that it is the expansion of life that has made it so different. For example, oxygen and methane left alone react to form carbon dioxide and water. Yet these substances are maintained on earth in quantities incommensurable with those that would be obtained if the system tended towards thermodynamic equilibrium, and in quantities far greater than those that would be obtained if only biotic factors were taken into account. This is due to the combined role of archaea, on the one hand, which produce methane under anaerobic conditions (without consuming oxygen), and plants, bacteria and phytoplankton, on the other, which produce dioxygen by photosynthesis.[23] Thus coupled, these two biotic sources supply the system with methane while exerting a negative feedback on it by reusing a large part of the CO_2 generated. By comparison, the atmosphere of Venus is made up of more than 98% CO_2 and that of Mars of 95% of this same gas. Without the intervention of living organisms, the earth's atmosphere would have a similar composition to that of those two planets. Yet it contains only 0.04% CO_2.

It is therefore unreasonable to explain life as an emergence of complexity against a background of pre-existing physical-chemical and geological conditions. This is broadly true, but only broadly. It forgets that life produces and shapes the environment – mineral, geological, atmospheric and climatic – that enables it to sustain itself. One of the philosophical lessons to be learned from the biographies of carbon is precisely that the question of emergence is badly framed. A key aspect of life is missed if it is explained by starting with the non-living, because the environment is not already a given. But carbon ignores the boundaries between mineral and living things, the chemical and the biological. It constitutes both at the same time and their interaction enriches diversity on both sides.

Carbon challenges the pictures of genesis set up by the three puzzles this chapter has touched upon: (1) the formation of carbon in stars from primordial nucleons; (2) the formation of life on earth from carbon; and (3) the stabilization of the climate from its biotic and mineral components. In each case, we tend to think of the emergence of a whole that is superior and irreducible to the parts from which it is derived. However, carbon teaches us that in a sense *the parts are greater than the whole.*[24] Carbon won't be enclosed within any totality; if it is, it escapes immediately. Never confined to one kingdom, it circulates endlessly among minerals, plants and animals. It ignores the strict level hierarchies that are supposed to constitute the ontological foundation of our earthly world. It lets itself associate with anything that can grasp its affordances. It is not enough for carbon to exist for life to develop; life had to be world-creative with carbon. It took the work of myriads of micro-organisms for the planet earth to become an ecumene, a durably inhabited and habitable land.

16

Turbulence in the biosphere

Carbon has been the talk of the town for some decades now, in scientific circles as well as in industrial, agricultural, economic and political ones, where it has provoked a general movement against it. War has been declared. All the talk is about 'fixing' carbon, 'sequestering' it in sinks or reservoirs, because it is poisoning the atmosphere. Circulate or fix, that seems to be the alternative. Could carbon have only two modes of existence, one sedentary, the other migratory?

Redox carbon

After a long, quiet existence of a few million years, during which it was fixed in sedimentary rocks, carbon began a more turbulent existence, constantly circulating around the air, land and sea, as soon as life appeared on planet earth. Its quiet existence was due to the virtuous circle of the thermostat effect of CO_2 degradation, which stabilizes the earth's climate when it tends to get out of control.[1] This feedback mechanism has ensured the stability of the carbon cycle for millions of years, even if the thermostat is set higher or lower depending on the period. However, as it takes hundreds of thousands of years to adjust this thermostat, we cannot rely on it to limit global warming to two degrees Celsius by 2050.

As for the carbon cycle in the biosphere, it works, *on the face of it*, like a fine-tuned machine. The atmosphere, which contains a very small fraction of the total quantity of carbon (even when it is in excess and accentuates the greenhouse effect), is above all a place of transit, of exchanges of carbon between land and ocean. For several decades now, the atmosphere has been closely monitored. Every day, the concentration of carbon dioxide is measured in parts per million (ppm) at a dozen sites around the world. For other environments, the quantities of carbon are measured in billions of tonnes or gigatonnes (GtC) (Table 2). The overall

Table 2. Concentration of carbon in the biosphere.
Atmosphere: 800 GtC
Ocean: 40,000 GtC
Soil (rocks and sediments): 100,000,000 GtC
Biomass: 600 GtC

volume of exchanges among atmosphere, land and sea is estimated at 90 GtC/year. The ocean absorbs around 2 GtC/year.

The proportion of carbon stored in the ocean is fairly easy to assess overall, as it is a more homogeneous medium than land. It is estimated that the ocean contains sixty-three times more carbon than the atmosphere. It stores a small proportion of this in organic form (micro-plankton) in surface waters, and the majority (over 50 million GtC) as a mineral, calcium carbonate, on the seabed.

The terrestrial biosphere is the largest reservoir of carbon, although it is difficult to quantify reserves because of the heterogeneity of soils and environments. Carbon is found either in organic form in plants (grass-lands and forests) or in soils. Peat, made up of plant debris, is estimated to contain 1,600 GtC, twice as much carbon as living biomass.

The carbon cycle is attracting increasing attention from researchers in all disciplines because it both determines and depends on the earth's climate. Carbon dioxide released into the atmosphere, along with methane, produces the notorious greenhouse effect that warms the climate. Hence the rush to 'fix' or 'sequester' carbon. This way of framing the problem has one advantage. Although it seems difficult to regulate the movement of carbon like a border security force, we can hope that by applying the good old method of balances, inherited from Lavoisier's chemistry, we will manage to equalize inputs and outputs. This assumes, of course, that the total quantity of carbon in the earth system is fixed.

The carbon cycle among land, ocean and atmosphere involves two main types of chemical combination: oxidation and reduction. In the oxidized state, carbon is characterized by an excess of positive charges. In CO_2, for example, a C atom has 'donated' two electrons (negatively charged) to each of the oxygen atoms, and is in the +4 oxidation state. In its reduced state, on the other hand, carbon has a negative charge. In methane, CH_4, for example, each hydrogen atom 'donates' an electron and the carbon atom has an oxidation state of -4. When oxidized, carbon forms inorganic compounds such as carbon dioxide, CO_2, or calcium carbonate, $CaCO_3$; when reduced, it forms organic compounds such as methane, CH_4. Fossil hydrocarbons, from wood to anthracite and oil, correspond to reduced carbon stocks. So we are talking about *reducing* carbon. But let's be clear: 'reducing' does not mean reducing the quantity of carbon on earth, which makes no sense. It means reducing in the redox sense, i.e. lowering the degree of oxidation of carbon. So the urgent need is not so much to 'decarbonize' the economy, as they say, but rather to manage the relative amounts of organic and oxidized carbon.

But this very static way of posing the problem is deceptively simple. It does not take into account the processes of association that take place in an interconnected system where actions are amplified by living organisms. Carbon compounds all have their own kinds of behaviour, qualities and rhythms, forming a complex and unpredictable system. They do not fit neatly into the static vision of flows between reservoirs that has inspired chemical balances.

Selfish carbon?

In 1976, Richard Dawkins put forward a personal and controversial theory of evolution, according to which natural selection applies to the genes that replicate living things and not to individuals or species.[2] Genes would thus be at the controls of the individuals bearing them. Organisms were there to serve the reproduction of genes, and the survival of DNA.

If Dawkins coined the 'selfish gene' concept to present the point of view of the gene that uses organisms to deploy itself through time, could we not imagine carbon as a malicious being that has infiltrated living organisms in the form of a DNA molecule, taking advantage of the genes themselves and the evolutionary processes to extend itself and spread across the planet? In fact, not only does every living thing carry a code written in DNA's carbon skeleton, but through their evolution, living things allow carbon to circulate and occupy all environments . . . to the point where today it has the unhappy honour of having the role of the lead actor on which the future of the planet depends.

Carbon regulates the material economy of the biosphere. As the nineteenth-century chemist Jean-Baptiste Dumas put it: 'From the chemical point of view, animals constitute veritable combustion apparatuses, by means of which carbon is constantly burnt and returns to the atmosphere in the form of carbonic acid', while 'plants, in their normal life, decompose carbonic acid in order to fix the carbon and release the oxygen'.[3] From this 'chemical statics' perspective, living things seem to be mere stopping-off points for carbon racing through the three kingdoms of nature. Carbon takes a little rest in the plant kingdom, is absorbed for a while in the animal kingdom, then resumes its mad dash through the atmosphere thanks to animal respiration.

Although fanciful, the image of carbon as a selfish exploiter of living things has the merit of being educational. The literary process masterfully employed in the carbon chapter of Primo Levi's *Periodic Table* mentioned above is repeated by Eric Roston in *The Carbon Age* and by Jan Zalasiewicz in *The Planet in a Pebble*.[4] By following the fate of the atoms in a pebble, treated as a microcosm of the universe, Zalasiewicz brings off a *tour de force*, travelling from the great cycles of the beginnings of the universe to the current ecological crisis.

Stories of this kind make an effective contribution to disseminating knowledge to the public about the mechanisms of global warming. They do, however, convey a skewed view of the problems associated with carbon. With carbon, as with genes, these authors essentialize an entity, more or less personified, which becomes the hero of the adventure. To the genetic ontology of biologists corresponds an atomic ontology among climatologists. They are interested in a stable identity that is preserved by transmission or transformation. But carbon, like genes, is not selfish in the sense that it behaves in such a way as to increase its chances of preservation at the expense of other elements. It evolves in an environment where its presence and circulation depend on its interactions with other elements and other systems. As we have seen above,[5] carbon is essentially a 'relational being': it offers a wide range of bonding possibilities that nature plays with, just like chemists constructing molecular structures. And this game mobilizes multiple interaction and feedback processes that further invalidate an atomistic ontology focused on a specific entity, carbon.

The carbon cycle in the biosphere illustrates this relational identity on a large scale, because it is a complex system, closely interdependent with two other regulatory systems – that of oxygen and that of the pH (measure of acidity) of the ocean – which regulate its circulation in the biosphere. Oxygen, produced by plant photosynthesis, is present in the earth's atmosphere and conditions everything: a drop in concentration would threaten living organisms and a rise would cause everything to burn. James Lovelock compared it to the 'battery charge' of the entire biosphere when he introduced the Gaia hypothesis, stressing that the charge level must be adjusted to the nearest tenth.[6] The concentration of oxygen in the different environments of the biosphere is a function of the respective proportions of organic carbon and oxidized carbon. And conversely, the carbon cycle is closely linked to the quantity of oxygen available in each of these environments. As for the pH of the ocean, it depends on the flow of calcium carbonate (which behaves like a base). The input comes from the dissolution of limestone rocks and the output from the calcium carbonate that is deposited as sediments on the seabed.

A melting pot

The biosphere thus appears to be a place of constant, intertwined movements between different milieus. Tyler Volk, professor of environmental sciences in New York and author of several fascinating books on the evolution of the earth, compares the biosphere to a gigantic stew, where encounters and mixtures take place under the influence of agitators such as winds in the air, waves in the ocean and worms in the soil.[7]

Plunged into this cauldron of movements and jolts, we have the choice of two possible attitudes. One is to try to correct the trajectories, to act on the climate by causing rain, snow, heat or cold. This attitude seemed normal and reasonable to those who believed that our mastery of nature had no limits and that engineers could make the desert green again. The great powers would have to assert their power even more by modifying the rainfall regime, and the United States clearly understood the strategic stakes of such modifications by carrying out various trials during the Vietnam War. In the context of the great plans for transforming nature during the Stalin era, the climatologist Mikhail Budyko declared that 'it becomes incumbent on us to develop a plan for climate modification that will maintain existing climate conditions, in spite of the tendency toward a temperature increase due to man's economic activity'.[8] A first official report[9] submitted to the US government in 1965 mentioned the risk of climate disruption as a result of CO_2 emissions due to the massive use of fossil fuels, and proposed counteracting these changes by geoengineering methods without considering the reduction of CO_2 emissions. And in 1992, the National Academy of Science again proposed taking action through geoengineering.[10]

In the face of ambitions or temptations to control and master all the flows of nature, a more modest attitude is to try to mitigate the effects of excess CO_2 in the atmosphere. To do this, we first need to understand what's going on in the machinery. What are the cogs and relays that make the carbon cycle work? So, let's take a look at where and how carbon enters the oceans and soils and combines and fixes itself there.

Star of the oceans: Emiliana huxleyi

Emiliana huxleyi is the name[11] given to a pelagic phytoplankton microalga that is widespread from tropical seas to the North Sea. This single-celled organism, around five microns in diameter, is one of the main members of the coccolithophore family, so named because they are enveloped in a protective white shell that lets light through (from the Greek *kokkos*: white, *lithos*: stone and *phorein*: to carry). The *E. huxleyi* coccosphere is a delight for researchers on materials, who marvel at its structure. It is made up of around fifteen curved slabs placed side by side. Each plate, or coccolith, is composed of a double crown, consisting of forty or so segments of calcite ($CaCO_3$), all curved to fit the shape of the cell (Figure 15).

While *E. huxleyi's* robust yet lightweight structure is attracting the attention of materials scientists who are trying to understand and learn about the process of biomineralization, climate scientists are interested in the climate-regulating role of coccolithophores in general. In their view, these calcite gems are 'biological pumps'. In fact, to build their calcium carbonate skeleton, microalgae capture carbon dissolved in water and, like all plants, release oxygen. The more CO_2 there is dissolved in the surface layer of marine waters, the more they proliferate. They are sometimes so abundant that 'blooms' appear – long, opaque, turquoise-white streaks mistaken for 'ocean snow' – floating at the high water mark.[12]

Coccolithophores do not always float between two water levels. During their lives, they continually peel off and drop their coccoliths

Figure 15. *Emiliana huxleyi* coccolithospheres.

onto the seabed. These sediment, fossilize and massify. In this way, they constitute a substantial, sustainable carbon sink.[13] These shells feed bacteria that transform them, creating chalky sedimentary layers.[14] After a few million years, they form the beautiful chalk cliffs of Dover or Étretat, characteristic of the Cretaceous, or the chalk age (Figure 16).

These tiny coccolithophores, of which several hundred species have been identified, are powerful allies in the fight against the greenhouse effect. Their current proliferation is both a symptom of the acidification of the oceans, due to excess CO_2, and a remedy, since it enables carbon to be stored and confined to the seabed. So we would be delighted if carbon circulation were the only factor. But selfish carbon is just an illusion, disproved by the indirect effects of this abundant carbon fixation activity. In the oceanic system, where everything is inter-dependent, the proliferation of coccolithophores is at the expense of other microalgae such as diatoms, which fix silica. This leads to the disappearance of certain zooplankton that are unable to consume the coccoliths, resulting in a lack of food for cod. As well as threatening animal biodiversity, the proliferation of coccolithophores disrupts the climate because, by absorbing CO_2, coccoliths form dimethylsulphide, which creates clouds and tropical storms. Finally, the curative effect

Figure 16. The Étretat cliffs.

of coccolithophores can only be limited because, in cold waters, the acidification of the oceans makes it more difficult to produce calcium carbonate. In short, the beautiful *Emiliana* alone cannot save the planet.

The potential of soils

For thousands of years, farmers on every continent have taken care to use a variety of techniques to maintain the soil. For example, terraces on the sloping soils of the Mediterranean basin help to combat the erosion that constantly threatens them. Manure has been used regularly to fix organic carbon in the soil. In this way, agriculture has become a powerful means of helping to fix organic carbon not only in plants but also in soils.

Industrial agriculture, however, which is all about improving plants, spread fertilizers *en masse* to increase yields, but to the detriment of soil fertility. The 'green revolution' has radically questioned this vision. Agriculture is currently responsible for 24% of greenhouse gas emissions (carbon dioxide, methane and nitrous oxide).[15] The main emission factor is farming ruminants, followed by synthetic fertilizers and pesticides, then rice paddies, manure spreading, tractors and farm machinery. And the rate of emissions is growing in proportion to the progress of agriculture in emerging countries. The carbon stored in the biomass produced by agriculture does not offset emissions because this biomass is consumed too quickly, sometimes burnt as a biofuel.

Storing carbon in biomass to the tune of massive reforestation campaigns therefore seems an urgent priority. Trees reduce carbon in the atmosphere through photosynthesis and store it in organic form for years. Reforestation is therefore an effective method of climate engineering.[16] However, the hoped-for benefits need to be tempered: on the one hand, a forest emits CO_2 when the leaves decompose on the ground; and, on the other, the forest has biophysical effects on the environment – modification of the albedo, evapotranspiration – likely to lead to local warming, particularly in arid regions.

Then there is the soil. It is a gigantic carbon sink that absorbs around two GtC per year, even if the storage processes are not always well understood. Carbon content peaks at 20 cm and then decreases exponentially with depth. The carbon sequestered in the soil helps to retain nutrients and encourages the growth of micro-organisms.

Soil is not only a support for biomass production, it is also a habitat for a host of organisms that recycle carbon and influence soil texture. These 'soil engineers' are extremely diverse, from worms to amoebae.[17] Worms, to which Darwin dedicated his last book,[18] have a well-studied role as mixers of organic and mineral matter in soil formation, but the functions of micro-organisms are more difficult to pin down. Genomic sequencing has shown that a single gram of soil can host up to ten thousand different bacterial species and almost a billion bacteria. Little is known about the respective contributions of bacteria, archaea and fungi, but overall they perform several essential functions: the biodegradation of organic matter, the production of nutrients for plants, nitrogen fixation and the degradation of pollutants (toxic metals in mining environments, for example). They play such a fundamental role that the carbon and nitrogen cycles in soils are almost entirely dependent on these micro-organisms.

The level of carbon in the soil depends on the ratio between the input of fresh organic matter, rich in organic carbon, and the old compounds mineralized by bacteria and fungi. In the Paris Basin, an estimated 50%

of soil carbon has been lost in fifty years as a result of the intensification of agriculture, the use of synthetic fertilizers and pesticides, and deep ploughing that oxidizes carbon. In an attempt to halt the rapid decline in soil 'carbon capital', a number of farming techniques are being tested. With sowing under plant cover, one crop is dedicated to food and the other to rebuilding the soil's carbon capital. Agroforestry, which combines trees and crops on the same plot of land, significantly increases biomass production through multiple synergies. Trees, in particular, increase its fertility since each year they return 40% of their biomass to the soil as their roots help to structure it.

More generally, what can be done to restore the soil's potential as a carbon sink?[19] Several solutions are being studied, all of which demonstrate the diversity of ways in which carbon exists, as well as the difficulty of achieving the balance of inputs and outputs recommended by agricultural chemists.[20] The simplest solution is to use carbon-rich inputs to provide organic carbon.[21] This is why, as early as the seventeenth and eighteenth centuries, scientists such as Duhamel du Monceau recommended fertilizing fields with coal.[22] Today, the addition of biochar (produced by carbonization, like charcoal) is once again being advocated to halt the alarming decline in soil carbon content. Adding biochar seems to be a win-win solution: fertility is increased – so more fresh organic matter is produced – and the carbonization of plant biomass can provide energy instead of fossil fuels.

However, increasing the quantity of carbon is not enough. It must also have sufficient residence time in the soil to increase fertility and, consequently, biomass production. Compost must also be added to the biochar, as soil quality is highly dependent on the input of fresh organic matter.[23] Not all organic carbons are equal, because they do not have the same interaction potential. Carbon can remain in soils for centuries because it resists degradation by micro-organisms. On the other hand, fresh organic matter, rich in cellulose, degrades more quickly. The respective proportions of labile organic matter (easily transformed) and recalcitrant matter (difficult to decompose) influence the dynamics of the carbon cycle. Inputs of fresh or labile organic matter significantly alter the rate of degradation of recalcitrant matter. Sometimes these inputs slow down the mineralization process, and sometimes, and this is the most frequent case, they galvanize it. The addition of organic matter occasionally tends to increase or even double the rate of carbon mineralization.

This effect, known as priming, has long been observed in soils and is now the subject of a great deal of research, because of its impact on the soil carbon budget.[24] The mechanisms at work, involving the production of enzymes by microbes, are still poorly understood and are the subject of much debate.[25] The effect seems to be more pronounced in certain places and at certain times, particularly when leaves are falling heavily in autumn. But this kind of synergy between two types of organic matter, with potentiating or inhibiting effects, seems to be a general phenomenon that does not depend on soil type, ecosystem or latitude.[26] It can occur in aquatic environments. In particular, it could be involved in the eutrophication of waters as a result of excessive nutrient inputs that cause algae proliferation and the death of fish due to a lack of oxygen in the water. In the current global context, the priming effect can be catastrophic because a high concentration of CO_2 in the atmosphere increases the mass of plant residues entering the soil, accelerating the carbon cycle in the soil and therefore the removal of mineral carbon. This general phenomenon sets up a vicious circle: the increase in CO_2 in the atmosphere intensifies the priming effect, which accelerates the removal of CO_2.

If carbon is not selfish, it can at least be described as capricious. Its cycle in the biosphere certainly lends itself to the drawing up of balances. Balancing inputs and outputs, managing the 'carbon budget' like a good accountant, is undoubtedly a more reasoned approach than using geoengineering to combat climate disruption. But these carbon budgets are based on a very substantial conception of carbon, which in all cases profoundly ignores its 'associative life'.

Carbon does not *circulate* like a vehicle on the public highway, with its predefined routes and codes. Rather, it follows a trajectory of local and contingent associations, whose partners are beings of very diverse natures, such as phytoplankton or micro-organisms. It is therefore high time to move beyond the current 'carbon politics', which treats carbon as a migrating, circulating population that must be counted and duly recorded in order to be controlled. There is an urgent need to develop knowledge that is attentive to the phenomena of synergy and feedback loops that produce unruly non-linear effects. Each form of carbon has its own ways of interacting with the environment, which determine both its mode of existence and its lifetime. Chemical statics offer only a slight insight into the complex dynamics of the carbon cycle in the biosphere. Adjusting the balance between organic and oxidized carbon depends on how the time factor is taken into account.

17

Rethinking time with carbon

Following the paths taken by carbon since its formation in interstellar space, its long sojourns in the lithosphere and its ceaseless journeys through the biosphere, it might seem as if the sequence of the chapters in this volume was unravelling the thread of time. Such a journey underpins the development of the concept of the Anthropocene, which is largely based on data concerning the circulation of carbon between the various spheres of the earth system.

The Anthropocene

This term, introduced by chemists Paul Crutzen and Eugene Stoermer in 2000,[1] refers to a new epoch in earth's history that began when human activities had a global impact on the earth system. Removals and discharges due to industry, agriculture, fishing and urbanization have reached such a level that they now appear to be greater than natural flows and are significantly disrupting the major balances of the earth's ecosystem. The Anthropocene follows on from the Holocene, which began ten thousand years ago, making the Holocene an unusually short period on earth – the Anthropocene may prove to be even shorter.[2]

An era in geological history is first and foremost a convention, the result of an official decision and agreement. The Anthropocene has been duly validated by an international stratigraphic commission,[3] but has yet to be ratified by the International Union of Geological Sciences. As far as the carbon cycle is concerned, the basic data are known and widely disseminated. The concentration of CO_2 in the atmosphere was around 260 ppm ten thousand years ago at the start of the Holocene, and 280 ppm in 1750. It is currently increasing by around 2 ppm a year, and passed the 400 ppm threshold in April 2014, a situation unseen for three million years.

It is worth pointing out that isotopic tracers, particularly ^{14}C and ^{13}C, are used to distinguish between anthropogenic carbon and carbon from natural flows. Like the mephitic gas of the ancients,[4] carbon is both problem and solution, poison and remedy. Thanks to isotope traces found in sediments and ice cores, we have been able to establish that 73% of the carbon released into the atmosphere comes from the combustion of fossil fuels, 25% from deforestation and 2% from cement works.[5] The quantity of carbon released into the atmosphere by human activities is estimated at 6.4 GtC/year in 1990, 7.2 GtC/year between 2000 and 2005 and around 10 GtC/year in 2015, out of a total estimated flux of 100 GtC/year.[6] During the 1990s, the ocean absorbed around 180 GtC, or a third of the emissions caused by human activities. The earth absorbs around 2 GtC every year. In principle, the natural carbon cycle is capable of responding to our disturbances, but over millions of years.

The central issue raised by the notion of the Anthropocene is the acceleration in greenhouse gas emissions since the industrial revolution, as measured by greenhouse gas concentrations in ice sheet records.[7] This is why Crutzen first placed the start of the Anthropocene in the 1780s on the basis of CO_2 fluxes and demographic data.[8] As the carbon cycle is strongly conditioned by the climate, and influences it in return, researchers sounded the alarm a few decades ago, and today a large number of them are active within the IPCC (Intergovernmental Panel on Climate Change).

The Anthropocene has thus become as much a political and social issue as a natural science concept.[9] As early as 1970, the futurologist Alvin Toffler emphasized the stress and disorientation caused by the flow of innovations and over-rapid change.[10] The German social theorist Hartmut Rosa has described a convergence of accelerations – technological, political and social – and concluded that time itself is in crisis.[11] The massive spread of digital technologies has further reinforced the importance of acceleration, which stands out as a dominant characteristic of our times.[12]

Hence the success of the Anthropocene concept, which has spread rapidly through society and the media, where it is fuelling controversy. While the debate on the nature of the Anthropocene (geological age or epoch?) mainly concerns geologists,[13] discussions on the date of the beginning of the Anthropocene already include ecologists, geographers and historians.[14] Clearly, dating the Anthropocene back ten thousand years or to 1945 does not have the same political significance insofar as responsibility is attributed to different actors.[15] As for the controversy over the name, it involves researchers from all disciplines. For Patrick de Wever, a geologist at the Muséum National d'Histoire Naturelle in Paris, the term 'Poubellian' [from the French *poubelle*, meaning 'rubbish bin'] would be more appropriate to describe an era that future palaeontologists will identify above all by its waste, in particular the plastic soup in the oceans.[16] This suggestion echoes that of marine biologist Maurice Fontaine, who proposed the term 'Molysmocene' (age of waste) in the 1960s.[17] For sociologist Jason Moore, the term 'Anthropocene' conceals the inequalities between human beings and masks the collusion between industrial capitalism and the intensive exploitation of fossil fuels, such as coal and oil;[18] he therefore proposes the term 'Capitalocene'.[19] For historians and sociologists, invoking the human race as a whole contributes to depoliticizing the analysis of the ecological state of the planet, which requires more refined historical categories.[20] By unifying humanity as a telluric force, the concept of the Anthropocene overlooks the socio-cultural diversity and economic inequalities of human beings. Surprisingly, fewer people opt for the term 'technocene'. According to Victor Petit, even though the notion of the 'technosphere' is not new, the term 'technocene' has the advantage of drawing attention to the fact that technical development tends to turn the earth into a material, energy and information 'machine-world' (an earth 'that measures itself'), operating as a closed system – and therefore entropic.[21] But that doesn't solve everything, insofar as it is the living earth – the biosphere – that is concerned by the Anthropocene.

A grand narrative

Whatever the variations in names or dates, the Anthropocene has established itself as a grand narrative. This is paradoxical enough because, as Sébastien Dutreuil has shown, the Anthropocene is not strictly speaking a concept of earth history. It refers to processes of forcing, disturbance and feedback, in other words to the complex systems of an anthroposphere rather than to the instigation of a geological era.[22] But this narrative has a dramatization function that Jean-Baptiste Fressoz likens to the aesthetics of the sublime.[23] As in the sublime, it gives rise to awe and terror.

Yet this grand narrative places humans at the centre of the earth's history. Hence a second paradox: the concept of the Anthropocene produces an effect that is the opposite of what it refers to. Entering the Anthropocene means that humankind is acting as a natural force, since our industrial activities are having a dangerous impact on the biosphere's carbon cycle. The interweaving of human infrastructures with the mechanisms of nature is proof of our condition as earthlings: we are full members of the biosphere, just like the plants that store carbon and the animals that dissipate it. And we share the same destiny of inter-dependence: '[T]he Anthropocene represents a new phase in the history of both humankind and of the Earth, when natural forces and human forces became intertwined, so that the fate of one determines the fate of the other.'[24]

And yet the invocation of an *anthropos* subject, the human race, which is said to have accelerated the removal of carbon from the atmosphere to the point of affecting the life and geology of the planet, is a way of re-singularizing humans, of making them an exception in the system. Crutzen and Stoermer wrote in 2000:

Considering . . . the growing impacts of human activities on land and in the atmosphere, at all, including global, scales, it seems to us more appropriate to emphasize the central role of mankind in geology and ecology by proposing to use the term 'Anthropocene' for the current geological epoch.[25]

In this way, the Anthropocene reactivates the grand narratives typical of modernity, in which Humanity, supposedly alien to nature, is called upon to free itself from it by dominating it. It produces an image of the earth as a planet to be acted on, an object of management and engineering in our hands: the 'earth system'.[26] Since the planet has gone mad, we need to regain control of it, as if we were piloting a spaceship. Or, to be more precise, if the Anthropocene is updating the grand narrative of modern humanity, it is doing so in a disenchanted mode, by reversing the direction of the arrow of progress. We are heading for a 'catastrophe', in both the everyday and mathematical senses of the term. Collapse is inevitable, and it is an exponential, non-linear process that tips a system over the edge. By placing Humanity at the centre of world history, we are taking away all human power to act, and this powerless omnipotence is experienced as a destiny. Catherine and Raphaël Larrère highlight this collusion between the anthropocentrism of the Anthropocene and catastrophism:

> The seductive force of catastrophism is that it gives a meaning to history … that of a necessary collapse of the world as it is, and on the rubble of which the survivors will have to invent another way of living. Could catastrophism (and the related interpretation of the Anthropocene) be the new grand narrative of a disenchanted world? Has the announcement of a grim future awaiting us replaced the great promises of happy days ahead?[27]

This grand narrative resonates with archaic myths. In seeking to dominate nature, Man (always with a capital M) has released the carbon that Nature (capital N), in its wisdom, kept buried underground in the form of coal and oil. These fossil fuels are pure concentrated duration, fossilized organic matter, frozen, immobilized time. Now this slowly amassed capital, which has been lying hidden from the covetous, is like a devil that has been taken out of its box, out of the ground, to seep into the atmosphere. And we would now like to put it back so we can breathe again. The importance of the link between earth and sky lends theological overtones to certain commentaries on the Anthropocene, which are reminiscent of Prometheus' *hubris*, or the Christian version of sin. The human history of carbon is summed up in a myth of transgression and sin. A fault that must be redeemed by paying indulgences (carbon offsets) to save humanity.[28] Everything is played out between the powers above and the world below, between heaven and earth in both the literal and figurative sense. The question of carbon in the Anthropocene is invariably posed in terms of CO_2 sequestration in storage sinks. It's a matter of transferring carbon from the sky to the earth, or to the depths of the sea, from top to bottom, along a vertical axis.[29]

The accelerating arrow of time

Verticality is the only temporal dimension retained by the Anthropocene. It is true that vertical chronologies are an integral part of geological culture. By practising the art of deciphering the earth's past in rocks and fossils, geologists have got into the habit of visualizing time in superimposed layers, starting from the present and working down to the oldest foundations. For them, the earth is a buried cathedral. Geology developed as a field science in search of traces and fossils, postulating the invariance of the laws of nature over time. This postulate, known as uniformitarianism, makes it possible to connect an infinite number of past moments with our present on the same line. It is then possible to string events or rock formations together on a single chronological thread that are thousands or millions of years apart. The single timeline enables us to read a landscape by juggling powers of ten.[30] It also allows us to project ourselves into an unfathomable future, wondering what, from our era, will leave traces.

The concept of the Anthropocene is part of this tradition, with the addition of the great acceleration hypothesis, which leads to a telescoping of scales. The Anthropocene presupposes a single axis around which all the ages of the earth are threaded. And to visualize acceleration, geological eras are often seen in the form of a timeline or, better still, a spiral wound around a vertical axis, as in Figure 17.

The concept of the Anthropocene brings the image of a single timeline back to the fore. Far from calling into question the fundamental postulate of the geological time-scale, the debates on the nature (event, episode, or epoch) and the beginning of the Anthropocene reinforce the conviction that this scale is well founded. 'The geologic time-scale, in my view, is one of the greatest achievements of humanity,' concluded a geologist in an article devoted to this debate.[31]

The concept of a single, linear time is a powerful scientific tool. It is undoubtedly one of the jewels in the crown of modern science, a tool for understanding and prediction. The concept of a single, universal time serving as an external framework for all unfolding events was built up progressively in the context of urban and commercial life, then in Newtonian physics, before being enriched and clarified with the development of industrial capitalism, the railways, etc.[32] This concept of time – essentially Western – makes it possible to establish a purely quantitative common measure between lived time, historical time, biological time, geological time and cosmic time. To make all these times commensurable, to standardize things as heterogeneous as celestial revolutions, human history, the time-span lived by each individual and the frequency of a caesium atom, an enormous effort of abstraction was needed,

Figure 17. Vision of geological ages suggesting the acceleration of time over the last three million years (the tail of the spiral).

involving a number of instruments, calculations and social, economic and technical pressures. The construction of universal time is the result of a conceptual synthesis of natural processes – the alternation of day and night, the seasons, etc. – and religious or political conventions. As Norbert Elias rightly pointed out, it transcends the divide between nature and culture, between object and subject:

> Dating, and measuring time in general, cannot be understood on the basis of a conception of the world as split into 'subject' and 'object'. ... Not 'people' and 'nature' as two separate entities but 'people in nature', is the basic concept which is needed in order to understand 'time'. So the endeavour to discover the nature of time helps us to understand that the splitting of the world into 'nature', the sphere of the natural sciences, and human societies,

the sphere of the social and human sciences, presents an illusory picture of a split world which is the artificial product of an erroneous development within science.[33]

The combination of physical, technical, social and scientific infrastructures for counting, measuring and transporting time has shaped our experience of duration to such an extent that this single universal time seems natural to us. This hybrid concept is seen as an external, natural, objective and immutable framework. It enables us to order things that are totally heterogeneous along a single line and to build grand narratives running the course of deep time on the rungs of a ladder constructed by a few learned societies.

Unravelling the scales

The grand narratives that leapfrog over powers of ten to reach the Anthropocene – all out of breath because time, as they say, is speeding up – are very powerful. But they are also sources of paradox. While the Anthropocene redefines us as earthlings, underlining our belonging to the global ecosystem of the planet we inhabit, *anthropos*, with a wave of its magic wand, is suddenly perched on an overhanging pedestal to contemplate the passage of time. The vision of a single, linear time presupposes an external observer, embracing the future of the universe. Who, then, can enjoy such a point of view? A god? Or humans who still think of themselves as alien to nature, and see themselves as masters and possessors? Outside the earth in any case, since they are able to look down on its history. We could evoke the words of Henri Bergson, who castigated the claim of the intellect to grasp the whole of which it is only a part:

[E]volutionist philosophy … begins by showing us in the intellect a local effect of evolution, a flame, perhaps accidental, which lights up the coming and going of living things in the narrow passage open to their action; and lo! forgetting what it has just told us, it makes of this lantern glimmering in a tunnel a Sun which can illuminate the world.[34]

It is not easy to give up the privilege offered by rationality of being able to look down on all phenomena from a distance. Even Lucretius, a pioneer in the fight against anthropocentrism, could not resist the temptation to adopt the comfortable position of someone looking down on nature, suggesting an external fixed point from which to embrace the whole with his famous '*Suave mari magno*' at the beginning of Canto II of *De Rerum Natura*. "Tis sweet, when, down the mighty main, the winds/Roll up its waste of waters, from the land/To watch another's labouring anguish far.' The greatest sweetness, he writes, is to occupy 'the high/Serene plateaus, well fortressed by the wise/Whence thou may'st look below on other men/And see them ev'rywhere wand'ring, all dispersed/In their lone seeking for the road of life.'[35]

This external fixed viewpoint has been given a new lease of life with the blue planet, the image of the globe seen from a satellite, which has such seductive power in the socio-technical imaginary that it provides advertising slogans that bear no relation to its content.[36] In short, such narratives presuppose that humans are outside nature and enjoying an exceptional status, a presupposition characteristic of modernity and that the Anthropocene should precisely call into question.[37]

To unzip the grand narratives, nothing beats Donna Haraway's humour as she derides all attempts to define a new age on the arrow of time and cheerfully mocks all the alternatives proposed to the expression 'Anthropocene' by coining a barbaric neologism 'Chthulucene':[38]

'My' Chthulucene, even burdened with its problematic Greek-ish tendrils, entangles myriad temporalities and spatialities and myriad intra-active entities-in-assemblages – including the more-than-human, other-than-human, inhuman, and human-as-humus.[39]

To do away with the grand narratives, we need to question the very notion of a single, linear temporal framework, insensitive to the ways in which time passes and things move. The '-cene' of the Anthropocene is a scene, a stage for a representation that homogenizes the heterogeneous and links the multiple in the unity of the tragic. But let's not forget that this frame is a historically and socially constructed edifice that is not shared by all human cultures. And even in the West it is not self-evident.[40] Building time-scales makes it possible to compare all things, even if it means highlighting the bottomless number of zeros that separate them in time. The disparity between the durations of individuals or things is represented as a spatial distance on a scale. This same principle of commensurability applied to space forms the basis of carbon offset markets and carbon credits, which enable us to weigh up distant natural and industrial environments and claim to be able to balance gains and losses.[41] It lends itself to the commodification of nature, rethought as a set of ecosystem services.

Some ecologists and geographers are concerned about the spatial scales used in global studies because they lead to the decontextualization of all things – particularly biodiversity – in order to standardize them in a gigantic global database.[42] Services such as GoogleEarth (and soon GoogleSky) maintain the illusion of scale as a continuous homothetic circulation between the very small and the very large, whereas scale is the result of connections between a large number of snapshots located and linked together by a method of map projection. If scales are tinkered with, we can also unravel them, display their sutures and open up the beings caught in their meshes to other modes of composition, weaving links other than scalar hierarchies. Anthropologist Anna Lowenhaupt Tsing emphasizes that one has to ignore or deny local circumstances and historical contingencies to make knowledge and things scalable. She argues that scalability is an instrument of Western expansion and domination.[43]

An analogous criticism of 'scalalability' must be levelled at the great chronologies that embrace all ages from the fourteen billion years since the Big Bang, the three billion years from the appearance of photosynthesis to the appearance of *Homo* around two million years ago, in order to diagnose the present and foretell the future. This linear vision of time as a *monodrome* on which the various occupants of the earth system move only allows us to envisage variations in speed. Faced with acceleration, we call for slowing down – as demonstrated by the success of the slow food, slow city and slow market movements[44] – or even to stop growth, or to reverse it through degrowth.

Such an 'arrow of time' model is both comfortable, with orders of magnitude neatly nested together – from the ultimate constituents of the atom to mountains and volcanoes, from the Big Bang to the end of time – and frightening, because it leaves only a narrow margin for manoeuvre. The only possible solution is to reduce our consumption in order to reduce our carbon footprint. But given the billions of people on the planet, this seems rather compromised. So we're at an impasse as long as we think in terms of scale. We are forgetting that the biomass of humans on earth is smaller than the biomass of ants, even though their carbon footprint does not threaten the planet's equilibrium.[45]

It should be added that the current carbon footprint makes a mockery of the linear time measured by carbon-14. In fact, this measurement depends on the ratio of $^{14}C/C_{total}$ in the environment at the time of the organism's death. The method is based on the assumption that the ratio of carbon-14 to carbon-12 in the atmosphere is stable, at around 1.2%. However, over the last century, human activities have increased not only the level of CO_2 in the atmosphere but also that of ^{14}C. Following the Hiroshima and Nagasaki bombs, and then nuclear atmospheric tests, the ^{14}C level practically doubled in the 1960s. After the Chernobyl and Fukushima accidents, we were able to observe carbon-14 being captured and trapped by plants. It remains in the rings of trees for thousands of years. Admittedly, these effects are well known, and correction tables exist for dating according to the industrial revolution, post-bomb and post-eruption volcanic effects, and changes in the earth's magnetic field. But the biological and geological carbon cycles, linked to the state of the biomass and the absorption capacity of the oceans and soils, also cause fluctuations in the overall level of carbon in the biosphere, confounding the linear law of carbon-14 dating. In short, the Anthropocene is casting doubt on the direction of the arrow of time.

By measuring carbon-14, humans have learned to inscribe the traces of their activities along with those of other beings who lived and died on the same timeline; to inscribe their history and even their prehistory in carbon time, to write it in carbon. This linear time, punctuated by the decays of the ^{14}C nucleus, is the *physical* time of carbon. But carbon also inscribes many other temporalities.

Multiple temporalities

The carbon cycle invites us – perhaps forces us – to abandon this linear model. It calls into question both the uniqueness of time and its linearity.

Through the different combinations and chemistries to which it lends itself, carbon has been able to experiment with a wide variety of temporal regimes. In rock, it is associated for three hundred million years in the form CO_3, whereas in living organisms it renews its associations on average every five years. Quiet life and busy life are not just successive phases, stages in the life of carbon, as the great linear narratives would have us believe. They are modes of existence that coexist by virtue of the very duration of each of the mechanisms that regulate the carbon cycle. We mentioned in the previous chapter the interdependence of the carbon, oxygen and ocean pH cycles. However, each of these elements has its own temporal regime: we are talking about *thousands* of years to bring the pH of the oceans into equilibrium, whereas the thermostatic effect of CO_2 alteration takes place over *hundreds of thousands* of years. As for oxygen, it takes *two million* years to regulate its concentration in the atmosphere, thanks to a mechanism known as the homeostat.[46] Slowing down makes no sense, because all of these time-scales are beyond our control. Instead, we have to take into account the rhythm of each of the players in the cycle. To do this, we need to understand how the time regimes specific to each element, and to each mode of existence of the carbon element, adjust to each other to produce synergistic and regulatory effects.

The carbon cycle is in fact a composition of very heterogeneous temporalities. It brings together musicians who each have their own tempo. In this concert without a maestro, the individual scores for each instrument compose a music, a world that is more or less calm, more or less turbulent. Composing, even *composting* temporalities: the compost metaphor dear to Haraway ('I am a compost-ist, not a posthuman-ist: we are all compost, not posthuman'[47]) can be taken at its word when we talk about carbon. One only has to mention the cohabitation of two forms of organic carbon – fresh or labile biomass and biochar that is more resistant to decomposition – which has an impact on soil fertility and the soil carbon budget.[48]

Our linear vision of time, in which events follow one another in succession and each state of the system can be deduced from the previous one, does not sit well with carbon cycles in soils, the atmosphere and the oceans, which constantly involve positive or negative feedback effects and non-linear processes. The geophysicist David Archer has clearly emphasized that the global carbon cycle in fact conceals interlocking mechanisms that are not always well coordinated.[49] Not only do the major regulatory systems operate over very different time-scales, but they can also have antagonistic effects. As we have seen, the carbon cycle is essentially based on a highly beneficial feedback loop: rising temperatures increase the concentration of CO_2 in the atmosphere, as well as the rate of chemical reactions that break down the CO_2.[50] Carbon sequestration reduces the amount of CO_2 in the atmosphere, thereby lowering the temperature. This feedback has stabilized the earth's climate over millions of years, even if the thermostat has been set higher or lower depending on the period. The same mechanism can, however, cause highly destabilizing tipping-point effects. The ice ages that covered the Northern Hemisphere in ice were not simply the result of a change in the earth's orbit around the sun. Lower temperatures increase the solubility of CO_2 in water, which accelerates the sequestration of carbon in the ocean, which reduces the quantity of CO_2 in the atmosphere, which causes the temperature to fall . . . until the planet freezes. It is always the feedback loop between temperature and CO_2 degradation that sometimes stabilizes and regulates, sometimes amplifies the effects by triggering catastrophic processes. David Archer has an analogy:

> The funny thing about the carbon cycle is that the same carbon-cycle machinery both stabilizes the climate (on million-year time scales) and perturbs it (on glacial time scales), as if the carbon cycle were fighting with itself. It would be analogous to some erratic fault in the furnace, driving the house to warm up and cool down, while the thermostat tries to control the temperature of the house by regulating the furnace as best it can. Time to call the furnace guy![51]

In view of the different profiles adopted by carbon in the past, Archer wonders whether we are now dealing with the gentle carbon that buffers or the saboteur that amplifies. His answer is not very reassuring. With rising temperatures reducing the solubility of CO_2 in the oceans and therefore the storage capacity of this reservoir, and the thawing of the permafrost releasing huge quantities of peat and methane buried under the ice, we can expect the greenhouse effect to increase for several centuries. 'Perhaps,' writes Archer, 'carbon reservoirs are like TNT charges, and the burning of fossil fuels is like a violent detonator capable of causing otherwise stable ingredients to explode.'[52] There is no doubt, in any case, that we are dealing with emergent phenomena between constantly interacting elements, which defy our ability to predict and perhaps also our imagination.

What can we draw from these considerations on the carbon cycle in order to envisage a political response to the Anthropocene?

First of all, we have to send the apocalyptic and techno-optimistic discourses back to where they come from, because both are based on the premise of a single arrow of time. Given the trajectories that intersect, collide or harmonize to regulate the dynamics of the earth system, it seems unreasonable to believe that we can intervene to 'correct' the carbon cycle. But pitting geoengineering against palaeofatalism should not lead to a sceptical shrugging of the shoulders. Catastrophe is not only certain or imminent. It is *already here* and getting worse day by day. But it is not like the Last Judgment, the lifting of the veil on the truth. Nor is it *a* catastrophe – global, massive, unambiguous, staggering, Hollywood-style. In its concrete manifestations, it is much more than a break in the threshold or an effect of scale. The global, abstract and theological catastrophic visions blind us to the fact that disasters are not only chronic but even *polychronic*. Catastrophes are multi-factorial and multi-consequential, diffracted, Dantesque, protean and – why not? – 'chthuluiform'. The tragedy of the Anthropocene does not preclude a sense of the comic, which is not derision but rather the story of the little arrangements made by inhabitants and migrants of all kinds – human, nonhuman and composthuman – who create dynamics that are more or less stable or turbulent. The tragedy of the Anthropocene calls for a new 'human comedy', a narrative task of describing and inventing the *possibilities of life* in the Anthropocene.[53]

And then, given the importance and variety of ways in which carbon exists and acts in the earth system, and its incessant interactions and associations with other components, placing humankind at the centre seems rather out of place. Humans may be a geological, telluric and climatic force, but so are coccolithophores, bacteria and worms. Above all, the earth is not a 'centre' that could be occupied by any form of life developing its own X-centrism. The earth of earthlings is not the terrestrial *globe*, the fascinating 'blue planet' that the great myth of the Anthropocene offers up. The inhabited earth, the biosphere or the *ecumene* – if we wish to extend this notion to living things other than *anthropos* – is a thin film about a hundred kilometres thick, an interface that extends between the upper limits of the lithosphere and the stratosphere that has been named 'the critical zone'.[54] Following the associations of carbon means understanding it not just vertically, but in as many planes as there are loops on which it is imprinted; it means seeing it pass into the distance above and below, while at the same time drawing locally, loop after loop, the envelopes in which we live, with their soils, oceans, climates, breaths and respirations.

Carbon therefore invites us to break down global and globalizing images, starting with scales of space and time. Instead, we have to learn how to compose the different temporalities of carbon, maintaining their plurality *and* their potentials for interaction. We have to abandon both the predatory relationship with nature established by the intensive exploitation of buried carbon and its global critique, which makes us hate carbon. Rather, the Anthropocene confers a new status on humans. Instead of believing that they can dominate nature through technology, and regulate technology through ethics, humans should think of themselves as partners and live as such, heirs or table companions of a host of living and technical beings, and develop an awareness of the multiplicity of their modes of existence and coexistence, of their possibilities for being-together as well as their incompatibilities. Given the time-scales involved in the major regulatory mechanisms of the carbon cycle, we should be concerned not only about future generations of humans but also about future species that risk disappearing with them. Beings such as bacteria and coccolithophores have powers that our anthropocentrism makes us unaware of. They also persist in their existence for periods that are unique to them. And humans have to adapt their own ways of living and consuming to their rhythms, so as to create synergies rather than destruction.

As for the 'inanimate' things that are technical objects, they are part of life, and in return they disrupt it. It's time to stop treating them as our slaves and embrace them as beings-in-the-world who also have their own ways of fitting in, taking hold and opening up to their milieus.[55] Machines have their own metabolism, their own fragility, their own rhythms. We can, perhaps we should, emerge from the Capitalocene, if the climate forces us to do so. But whatever happens, we won't be able to emerge from the Technocene unless the earth emerges from it without us. It is essential not to separate the technologies of matter and the technologies of life, the life of technologies and the technologies of life – with carbon as the link.

In our view, the real challenge of political ecology is to recognize the active and multiple presence of carbon and the resulting collusion of temporalities. By paying close attention to the variety of ways in which carbon exists, the ways in which it resides or circulates, and the temporal regimes of its multiple associative lives, we can rethink our own ways of inhabiting the earth. Instead of sticking to the simplistic and largely mythical solution of bringing CO_2 down from 'up there' and burying it in carbon sinks that are not necessarily ready to receive it, it might be reasonable to focus on the surfaces and interfaces between the atmosphere and water or land. In these areas, which are richly populated with micro-organisms ready to break down carbon, there is some chance, provided we increase our knowledge, of mitigating the effects of CO_2. Make an alliance with microbes instead of relying on the genius of *anthropos*.

Epilogue
The heteronyms of carbon

The challenge of this book was to make carbon speak, to give voice to a being that is omnipresent in our environment, hidden in familiar objects and even in our bodies. At a time when technology and the economy aspire to dematerialization, it was worth highlighting the richness of our partnership with the material. Carbon, the hero of this essay, is a multiple being on which the fate of humanity largely depends. Firmly anchored in the material world, carbon tells whole swathes of our history, the history of the earth and of living things. It brings together the history of chemistry and technology, economics, geopolitics, ecology and anthropology.

In each of its states of oxidation, carbon has forged more or less durable links, creating its own arrangements and regimes of temporality. By revealing new affordances, each mode of carbon's existence unfurls a thick, rich history in which human, scientific, technical, industrial, capitalist and political adventures are played out. In recounting some of these adventures, this essay sets out some of the many alliances that Western societies have forged with various modes of carbon existence. These are vital links, so ubiquitous that they suggest that we owe carbon a huge debt in the current context of the commodification of nature. But these carbon-based narratives are also an invitation to show how natural and cultural, human and planetary histories are intertwined, and how we share a common destiny with carbon.

Stories of genius

This is why it is not possible to tell the story of carbon within the framework of the categories firmly established in our culture, such as the division between subject and object, between animate and inanimate, between humans and non-humans. In his Nobel lecture on the discovery of fullerenes, Richard Smalley did not hesitate to speak of carbon as a 'genius':

[T]he discovery that garnered the Nobel Prize was the realization that carbon makes the truncated icosahedral molecule, and larger geodesic cages, all by itself. Carbon has wired within it, as part of its birthright ever since the beginning of this universe, the *genius* for spontaneously assembling into fullerenes.[1]

The metaphor of the genius inhabiting carbon is not just an animistic projection. It must be taken literally. Thanks to its ability to form bonds with itself, as with most of the elements in the periodic table, carbon is a being dedicated to the most diverse 'assemblages'. The multiple capacities deployed by carbon in various organic or inorganic assemblages force us to conjugate in the plural the most fundamental categories of our thinking and our relationship to the world.

Now that carbon is a major player in our history, the ontological hierarchy of beings, which posits an active human subject grappling with inert matter that is more or less passive or malleable, seems strangely impoverishing. Carbon reveals the agencies of material entities, stripped of all activity and quality for too long by the dominant mechanistic tradition.[2] The 'vibrating presence'[3] of matter, its 'vitality', is occasionally perceived and recognized in aesthetic experiences, often in fleeting moments. But carbon provides powerful arguments for any plea in favour of things. It illustrates the activity of materials not only through its recalcitrance (e.g. in the stubborn persistence of non-biodegradable plastic debris) but also through its affordances (by offering indefinite possibilities for diverse connections).[4] Carbon is thus more than an object thrown in front of a subject who condescends to look at it, more than an obstacle to human designs, it is much more a thing that constantly surprises us by the strangeness of its behaviour, as in graphene, which we are just discovering even though we have had it right under our noses for a long time.

If carbon is a good indicator of the active presence of material bodies, of their persistence of being in the world, which Spinoza called *conatus*, it is because it unfolds in the long duration of geological time as well as in the evolutionary time of living organisms, in historical time as well as in the biographical time of individual lives. More precisely, each carbon has its own tempo, and these different temporalities coexist in the same place. All carbons are con-temporary.

A plurality of modes of existence

Finally, what is carbon? Over the course of this essay, we have unfolded its multiple modes of existence: genius of place, vital and lethal spirit, gas, exemplary element, metaphysical substance, relational being, organic skeleton, solid material, evanescent surface, tracer, fossil memory, intoxicating fluid, improbable cosmogonic core, general equivalent, speculative product . . . Carbon is all these things and more. It consists in the multiplicity of its modes of existence, in its different ways of being an object. So to speak of carbon as a chemical element at the top of the fourteenth column of the periodic table, to speak of carbon as a diverse set of materials, or even to speak of carbon as fossil energy, is to signify different ways in which carbon is an object. It means being situated in different modes.

This pluralist perspective was suggested to us by the chemical concept of allotropy, which carbon masterfully illustrates by being the signature for graphite and diamond, two bodies with diametrically opposed properties. But carbon's multiplicity of modes of existence is in no way reducible to its multiplicity of allotropic forms. Allotropy only makes sense in *one* of the ontologies that carbon helps to write, the one that opposes the element to its phenomenal manifestations as a simple body.

Although crucial for the writing of chemistry textbooks, the definition of carbon as a chemical element is not *the* definition of carbon. It is just one of the ways in which it signs its name as an element, thus inscribing a substantialist ontology that posits it as a permanent substratum under-lying its various phenomenal manifestations.[5] But the hierarchy of scales of reality suggested by this metaphysical – or 'metachemical' – mode of existence of carbon is due to the specific way in which this mode tends to reduce others to its own. And carbon's other modes of being do the same, each in its own mode, like the carbon equivalent on the markets of environmental capitalism, which tends to make all of carbon's lives dissolve into financial speculation. Each mode of existence has an 'imperialist' tendency, trying to constitute itself as a paradigm for the others: the universal standard, the nanoworld, the plastic continent, the world oil order, the anthropic universe, the Anthropocene, the general equivalent and the global carbon debt are all 'carbon worlds' that are explored in this essay. Each mode of carbon existence claims to be 'worlding' according to its own pattern, imposing its own ontology, norms, symbols and tempo.

And yet, all these carbons coexist, in all con-tingency and con-tempo-raneity, for better or for worse. Chase one out the door and it comes back through the window. So modern names like 'carbon dioxide', coming in with reforms to the language of chemistry, have never quite dispelled the archaic connotations of the spirit of life and death associated with the ancient Mephitis, which is now being reinvented in discussions on climate. This is why it was necessary to unfold the constellation of carbon's modes of existence and, above all, to firmly maintain its plurality, resisting the imperialist temptation to deduce it from just one of them. In short, to resist ontology.

But it is precisely when taken in its ontological sense that the notion of 'mode of existence' has its weakest meaning. Indeed, this notion belongs first and foremost to the most classical ontology, which considers that *being* comes first, and *only then* the modalities of being, i.e. the different ways of predicating an attribute of the same subject. We are saying, on the contrary, that when carbon deploys several modes of existence, we take this concept in a strong sense, inspired more by Gilbert Simondon[6] and above all by Étienne Souriau, who, in *The Different Modes of Existence*, proposes a 'multirealism' that is, to say the least, vertiginous.[7] Bruno Latour puts it this way:

> The key to this project is that he wants to be able to differentiate the modes of being themselves, not just the various different ways of saying something about a given being. The notion of modes is as old as philosophy itself, but up until now one's discursive orientation on the problem was that the modus was a modification of the dictum, which had the special status of remaining precisely the same as itself. In the series of phrases: 'he dances', 'he wants to dance', 'he would really like to be able to dance', 'he would so like to know how to dance', the 'dance' doesn't change. ... At first philosophers used this discursive model for the modalization of being by, for example, varying the degree of existence from potential to actual, but without ever going so far as to modalize whatever it was that went into the act. Predicates might be numerous and they might wander far afield, but they would always come back to nestle in their pigeonholes, in the same old dovecote of substance. ... Now, multirealism would like to explore rather different modes of existence

than the sole action of saying several things about the same being. Its whole aim is that there be several ways of being.[8]

In this strong sense, the concept of modes of existence is not only an invitation to epistemic pluralism – to a form of tolerance for the multiplicity of both scientific and everyday knowledge – but also an invitation to a pluralization of ontologies. Modalization attributes another way of being to what it is modifying, another ontology. Not only can carbon be spoken about in several ways, it *is* carbon in several ways. It is a multireal object.

However, we are using the concept of modes of existence in a slightly different way to Latour in his *An Inquiry into the Modes of Existence*.[9] The aim is not to propose a new metaphysics or a new ontology, even a pluralist one. The aim is not speculative. It is not to show how carbon revolutionizes the general science of being, but to propose another way of approaching the making of ontologies through writing, through narrative. It is less a question of talking about the *reality* of things than of making real *things* talk, things that express their existence in multiple ways, and letting their ways of being be written through us, we who *narrate* them.

Ontography

By describing carbon's various modes of existence, this essay has sketched out an 'ontography' of carbon. This term is to ontology what ethnography is to ethnology. It is a written description, a narrative that does not have the pretensions of a metaphysical discourse devoted to stating the essence or true nature of a thing. Ontography can also refer to a narrative form that is a mixture of biography and ontology; it is interested in the way in which things participate in the *writing* of ontologies. In this way, each of carbon's modes of existence typifies or proposes an ontology, a way for carbon to be inscribed in reality. Carbon as element signals a substantialist ontology, carbon compounds an ontology of dispositions, nanocarbons an ontology of affordances, all in their superficiality, with 'bodies without organs' and Deleuzian arrangements [*agencements*]. Ontography in no way postulates the necessity of the modes of existence it sets out to characterize, nor any hierarchy of levels of reality. Because it deflates the essentialist ambitions of ontology,[10] it lends itself to the deployment of relationships, without ever weaving a causal chain among beings or making deductions between levels. It lets beings define the levels of analysis relevant to their description.[11]

Ontography, as we understand it, differs from that proposed by those philosophers who promote 'speculative realism'.[12] Even if we share their orientation towards objects, we differ on what is meant by 'object'. The objects that speculative realism brings to the forefront of philosophy are mute objects, withdrawn into themselves, whose essence is removed from modes of access. They are objects for philosophers, who owe them nothing and which they do not have to describe well. Generally speaking, the ontography of speculative realists consists of a catalogue of being: first there is the being or the objects, then their cataloguing. Even Ian Bogost, who proposes a 'pragmatist speculative realism' with an ontography closer to our own, gives us a catalogue:

> Like a medieval bestiary, ontography can take the form of a compendium, a record of things juxtaposed to demonstrate their overlap and imply inter-action through collocation. The simplest approach to such recording is the *list*, a group of items loosely joined not by logic or power or use but by the gentle knot of the comma. Ontography is an aesthetic set theory, in which a particular configuration is celebrated merely on the basis of its existence.[13]

Instead of catalogues, we prefer books that tell stories. As our carbon-objects are beings-of-the-world that unfold a *vita activa*, their ontography is, rather, a series of narratives that make visible the contingent links they weave with various beings: how they associate, entangle their destinies, settle or move, disguise themselves and depart for other lives . . . These relational beings convey or give rise to values, and they engage our responsibility without being reduced to the relationships *we* have with them.

The aim is therefore to give voice, through stories, to the ways in which carbon exists. To do this, we use the metaphor of the carbon *writer*. What weight should we attach to this metaphor, and in what sense should we understand it to mean 'writing'?

Firstly, this metaphor has multiple threads. They may come from the practical or artistic uses of carbon as a writing tool (*graphein*), or as a tracing medium, inscriber, fingerprint fixer, archive and timekeeper; or from the intellectual acts it gives rise to in scientific culture as an instrument of taxonomy, tabulation or DNA and RNA encoding. Carbon is sometimes an artist, sometimes a scribe or an accountant; sometimes more text-like, at others a number or an image. The carbon-writer metaphor goes hand in hand with the refusal to divide text and image in a straightforward way (text being traditionally regarded by our culture as closer to *logos*, or speaking-reason). Numerous expressions such as 'photography', 'cinematography', 'echography', 'choreography' or 'scenography' associate the suffix '-graphy' with practices that involve images. *Graphein* does not separate writing from images, because it does not proceed from language, but rather gives rise to it and opens up its possibilities. It is in this sense that carbon writes and writes itself through us, and that we write our stories in carbon.

Secondly, 'writing' is understood here in a resolutely non-anthropocentric and non-logocentric sense. It is not the *logos*, the living word that precedes and invents writing. Writing is not the secondary translation of language, freezing it in signs. Quite simply, we are the ones narrating-reasoning. We are 'languaging' by linking (ourselves to) traces (*graphemata*), as we activate those traces. This notion of *languaging* forged by the cybernetic biologist Humberto Maturana[14] has now been taken up by the so-called 'enactive' and 'externalist' approaches to cognition, where we know not 'in our heads' but through active engagement and structural coupling with the environment. From this point of view, humans are not the only ones who language. All living things have this *languaging* capacity and it can even take place across several species. We can language with dogs, birds or cows without necessarily sharing *one* common language. But if this notion of *languaging* has brought something new to the cognitive sciences, it does nothing more than directly translate the Greek verb *logein*: not 'to say' (from a pre-existing language) or 'to speak' (of a pre-existing object) but 'to language', to bring language into being by activating a legacy (a *legein*) that opens up the possibility of *graphemata*. *Legein* and *logein* go hand in hand, but their possibility emerges from a more primitive *graphein*.[15]

Thirdly, the writing of carbon is to be understood in terms of the interplay between the two related notions of *graph* and *gram*. *Graphein* is all about writing, inscribing, noting, registering, marking, tracing, engraving, cutting, imaging and painting. *Gramma* means line, written sign, letter, trace, mark, inscription, record, character, small part or engram, i.e. memory trace. In French, the suffix *-gramme* is also found in the system of weights and measures,[16] for which carbon has been the standard of determination on an atomic scale since 1961. Graph and gram are both materials *and* images. The graph is a movement-image that has no content other than its own movement. Its materiality is gestural, and it may well last only as long as it is active, as in sign language (signing is graphing). Applied to a support, the graph inscribes a gram. The gram is an image-object. It is both the result, the imprint of the *graphein* gesture, and a resource for it, its material. The gram can be taken up again and updated in a writing gesture. As we showed in the case of the phono-graph/gramophone discussed in Chapter 9, the graph and the gram are reversible: we can move from gesture to sign and from sign to gesture without going back to the same thing, because writing is recursive, it multiplies possibilities. For example, let your finger run across the computer keyboard: 'nhgtreza'. What is engrammed is a singular gesture (characters randomly struck). But as soon as this gram fits into a system of pre-existing grams, even as contingent as the succession of letters on the keyboard ('qwerty'), a *grammar* appears. It then becomes possible to perform operations on the gram, to create anagrams through sequences of operations. The power of language to multiply the possible is derived from its graphic and traceological recursivity, not from its signifying function.

The trace always precedes the meaning. Writing is always also about reading traces, not fixing an oral thought that pre-exists it independently of any medium or tool for inscribing it. Similarly, reading involves trans-forming traces into language, so it is still writing (mentally or gesturally), or language-making. It is in this interplay between graph and gram that the carbon-writer metaphor must be understood.

Readers may be thinking that by going through the twists and turns of etymology we never get out of metaphors to reach the things themselves and rise to the level of pure *logos* that would allow us to contemplate them all. All we are doing is following the carbon trail from one thing to another: *meta-phorein*, 'to transport elsewhere'. They will be right, because if *logos* is languaging in the legacy of traces, then there is no way out of metaphor. It is irreducible. The best we can do is to play it as if it were a score, and let recurring themes be heard. For example, carbon as a general equivalent in its economic mode of existence plays a role analogous to that of the mole in the chemical mode of existence of carbon, that of a counting standard. But there is no ultimate, master metaphor, no meta-metaphor, only more or less effective metaphors involving narrative choices that engage biographers and make demands on them in return.

Who is carbon?

Primo Levi explains why carbon never stops telling stories:

> So it happens, therefore, that every element says something to someone (something different to each) like the mountain valleys or beaches visited in youth. One must perhaps make an exception for carbon, because it says everything to everyone, that is, it is not specific, in the same way that Adam is not specific as an ancestor. . . . And yet it is exactly to this carbon that I have an old debt, contracted during what for me were decisive days. To carbon, the element of life, my first literary dream was turned, insistently dreamed in an hour and a place when my life was not worth much: yes, I wanted to tell the story of an atom of carbon.[17]

Unlike the other chapters of *The Periodic Table*, each of which 'colours' a specific experience in the author's life by placing it under the sign of an element, the final chapter relates the experiences of a carbon atom writing its multiple lives through the author. For example, the first chapter, 'Argon' (solitary), is an opportunity for Levi to evoke his 'specific' origins: his ancestors, whom he knew little or nothing about, frozen in the inertia of death. In contrast, the final chapter, 'Carbon', associates the element of life with the common ancestor of all humans, Adam, who shares its etymology with diamond (adamant). The specificity of carbon is that it is not specific. It can change its mode of existence.

It was by thinking about carbon in the camps that Levi first felt the desire to write. It is his 'debt', his *legein*. And carbon literally *signs* the story of his life as a writer, putting the finishing touches to the book. Whereas the other elements say 'something different to each' and resonate with particular moments in the chemist's life, carbon does not speak of the idiosyncrasy of a person or the singularity of the author; it *brings him into existence* by making him a 'writer'. Carbon describes the practice of 'writing' in general. It only singularizes through commonality.

> I could tell innumerable other stories, and they would all be true: all literally true, in the nature of the transitions, in their order and data. The number of atoms is so great that one could always be found whose story coincides with any capriciously invented story.[18]

So there are an infinite number of possible true stories about carbon. Many have been told by popular science writers who personify the carbon atom or recount its tribulations from the dawn of time to the present day, moving from rocks to water, air, plants and animals. Mineralogist Simon Wellings claims to hear the voice of carbon:

> *A voice – soft and scratchy, like the sound of a pencil on paper – started speaking in my head. It claimed to be a Carbon atom and started telling me about the amazing journeys it's taken over billions of years. I started writing down what it was saying and I've added some of my own notes about the science (which seems correct). It must have been a dream though, surely?*
>
> I'm a Carbon atom. Carbon 12 to be precise, like most of us are. I've got 6 protons and 6 neutrons in my tidy little nucleus and usually some electrons buzzing around to keep me decent. I'm not one of these bloated Carbon 13 types, waddling around with their extra neutron. Still, they're OK, it's Carbon 14 I try to avoid – big fat lumps. So unstable too! I bonded with one once, you know, sharing some electrons, just chilling in a bit of soot. Then, bang! All of a sudden one of her protons shoots off and she's turned into a Nitrogen. Gave me such a shock.
>
> Oh, now, you're getting a bit worried by me talking to you, aren't you? Be careful, I'm only in your brain as a bit of glucose, not as one of your cells. If you think too much then two burly Oxygens will grab hold, stick me in your blood stream and puff me out via your lungs. My advice to you is this: don't worry about how I'm talking like this, just enjoy the story. You'll want to hear it. I've seen things you people wouldn't believe.
>
> So let's start at the very beginning; a very good place to start. Like most of us atoms, I was born in a star somewhere.[19]

Under the pen of biologist Fiona McCann, readers are hailed by carbon:

> Hi! My name is Carlos Carbon. I'm a carbon atom on the palm of your hand! I bet you're wondering how I even got here in the first place. It's quite a journey so let me tell you about it!
>
> It all started when I was just a simple carbon atom in the atmosphere! I combined with my friends Ollie Oxygen and Oliver Oxygen to become carbon dioxide.[20]

Children's books or blogs use these literary tropes with varying degrees of success.[21] Carbon lends itself easily to this kind of short story centred on a 'I, carbon' who explains to children who he is and what he does.

Stories that are all arbitrary and all true, all different and all a little the same, all anthropomorphic but not anthropocentric. Far from the arrogance of the grand narratives of the anthropic principle and the Anthropocene, which are currently using carbon to put *anthropos* back at the centre of our concerns, these small anthropomorphic stories help to decentralize and transport readers to the level of relations between objects; to show that there is no *single*, fundamental point of view that would account for all the identities of carbon, but a world to be composed from contingent tempos and points of view.

Finally, these stories have the merit of asking the right question. Not the philosopher's question: 'What is carbon?' *Ti esti?* But rather, *who* is carbon?

Just as the Portuguese poet Fernando Pessoa signed his works under several names, carbon has signed chapters in the history of the earth, of life and civilizations under several names. The term 'heteronym' coined by Pessoa to designate the fictitious authors who signed his works – Alberto Caeiro, Ricardo Reis, Alvaro de Campos, Bernardo Soares and some seventy others – does not refer to mere pseudonyms or pen names. Far from being the fictitious signatures of one and the same real author, heteronyms are the real signatures of fictitious authors, each possessing, to a greater or lesser degree, an imaginary life and his or her own voice, with a clearly identifiable style. Through his heteronyms, Pessoa deploys multiple personae and presents himself as one of many. Pessoa 'in person'[22] saw himself as the 'orthonymous' character of the author he was. While the term 'orthonym', which he applied to himself, connoted an upright, socially correct position, it did not suggest a deeper level of reality.

At the end of a long trek through millions of years, we are tempted to present the word 'carbon' as an orthonym, a signature, among others, that is scientifically correct but does not enjoy any ontological privilege. Mephitis has written its own story, all about places inhabited by obscure powers, while fixed air told us about the birth of the hunt for gases that were duly measured and characterized. Coal itself has many lives, as a fertilizer, as a fuel that fomented the industrial revolution or as a pollution filter. Oil tells the story of the rise of an empire whose end we can only guess at. Graphite inscribes its signature on the white sheet of paper before starting a new life in the atomic reactors and then giving birth to graphene, which is full of promise. As for the best-known heteronym, CO_2, it is creating its own market and the fate of both humans and ecosystems now seems to depend on it. These misadventures are just one of the many stories told by carbon, which has many others to its credit. Because carbon is above all a multiplicity, and speaks to everyone.

Notes

Prologue: Why write a biography of carbon?

1 The carbon–nitrogen–oxygen cycle is the main source of energy for stars that are the same size as or bigger than the sun.

2 Henri Bergson, *The Two Sources of Morality and Religion*, trans. R. Ashley Audra, Cloudesley Brereton and W. Horsfall Carter, London: Macmillan, 1935, p. 221.

3 Sam Kean, *The Disappearing Spoon: And Other True Tales of Madness, Love, and the History of the World from the Periodic Table of the Elements*, Boston: Little, Brown and Company, 2010.

4 Aristotle, *Politics*, trans. Benjamin Jowett, Kitchener, ON: Batoche Books, 1999, p. 158. Hannah Arendt represents the Aristotelian view of the human condition through the conjunction of a biological existence (*zoë*) and a political life (*bios*) which is engaged in language and action. *The Human Condition*, Second Edition, Chicago: University of Chicago Press, 1998 [1958].

5 Significantly, to celebrate the 150th anniversary of the discovery of the periodic system by Dmitri Mendeleev in 2019, the European Chemical Society (EuChemS) issued an unfamiliar table of elements illustrating the elements' scarcity (https://www.euchems.eu/euchems-periodic-table/). See also Bernadette Bensaude-Vincent, ed., *Between Nature and Society: Biographies of Materials*, Singapore: World Scientific Publishing, 2022.

6 Dimitris Papadopoulos, Maria Puig de la Bellacasa and Natasha Myers, eds, *Reactivating Elements: Chemistry, Ecology, Practice*, Durham, NC: Duke University Press, 2021, p. 5.

7 Jean-François Mouhot, *Des esclaves énergétiques: Réflexions sur le changement climatique*, Ceyzérieu: Éditions Champ Vallon, 2011.

8 This is the approach driving Eric Roston's book, *The Carbon Age: How Life's Core Element Has Become Civilization's Greatest Threat*, New York: Walker & Co., 2008.

9 In the case of graphite, a stack of hexagonal carbon atoms is bound together by weak van der Waals forces; in each of its sheets, called graphenes, each carbon atom is bound to three neighbours; for diamond, a face-centred cubic structure where each carbon atom is bonded to four neighbours.

234

10 The set of gases with the property of absorbing infrared radiation: carbonic acid, ozone, water vapour, methane, nitrous oxide, soot particles or black carbon, sulphur hexafluoride and numerous halocarbons such as chlorofluorocarbons.

11 Donald MacKenzie, 'Making things the same: Gases, emission rights and the politics of carbon markets', *Accounting, Organizations and Society*, 34/3–4, 2009, pp. 440–55.

12 Alain Gras, *Le Choix du feu: Aux origines de la crise climatique*, Paris: Fayard, 2007.

13 Primo Levi, *The Periodic Table*, trans. Raymond Rosenthal, New York: Schocken Books, 1984. The story of a number of atoms is also the approach taken by the geologist Jan Zalasiewicz in writing the story of a pebble: *The Planet in a Pebble: A Journey into Earth's Deep History*, Oxford: Oxford University Press, 2012.

14 Our carbon dioxide, mixed with some hydrogen sulphide fumes.

15 Étienne Souriau, *The Different Modes of Existence*, trans. Erik Beranek and Tim Howles, Minneapolis: University of Minnesota Press, 2015 [1943]; Bruno Latour, 'Reflections on Étienne Souriau's *Les différents modes d'existence*', trans. Stephen Muecke, in Graham Harman, Levi Bryant and Nick Srnicek, eds, *The Speculative Turn*, Melbourne: re.press, 2011, pp. 304–33; Bruno Latour, *An Inquiry into Modes of Existence: An Anthropology of the Moderns*, trans. Catherine Porter, Cambridge, MA: Harvard University Press, 2013.

Chapter 1 Mephitis

1 Dis is a Roman god of the underworld not unlike Pluto; Acheron is the underground river leading to the underworld; and the Furies are avenging deities.

2 Virgil, *The Aeneid*, trans. David Alexander West, London: Penguin Books, 1990, 7.563–71, p. 180.

3 Giovanni Ghiodini, Francesco Frondini and Francesco Ponziani, 'Deep structures and carbon dioxide degassing in central Italy', *Geothermics*, 24/1, 1995, pp. 81–94.

4 Giovanni Ghiodini et al., 'Non-volcanic CO_2 earth degassing: Case of Mefite d'Ansanto (Southern Apennines), Italy', *Geophysical Research Letters*, 37/11, 2010. DOI: 10.1029/2010GL042858.

5 Jens Soentgen, 'On the history and prehistory of CO_2', *Foundations of Chemistry*, 12/2, 2010, pp. 137–48.

6 On Roman, particularly Stoical, thought on breath and atmospheric ontology, see Emanuele Coccia's excellent book *The Life of Plants: A Metaphysics of Mixture*, trans. Dylan J. Montanari, Cambridge: Polity, 2018.

7 It is in order to discuss these kinds of questions that the historian of science Lorraine Daston suggested biographies of scientific objects. The biography

enables a reconciliation of realists and constructivists because it assumes that scientific objects are both real and at the same time historical. Lorraine Daston, ed., *Biographies of Scientific Objects*, Chicago: University of Chicago Press, 2000, pp. 2–4.

8 Michel Lejeune, 'Méfitis, déesse osque', *Comptes rendus des séances de l'Académie des Inscriptions et des Belles-Lettres*, 130/1, 1986, pp. 202–13.

9 Olivier de Cazanove, 'Le lieu de culte de Méfitis dans les Ampsanctiualles: Des sources documentaires hétérogènes', in Olivier de Cazanove and John Scheid, eds, *Sanctuaires et sources: Les sources documentaires et leurs limites dans la description des lieux de culte*, Paris: Publications du Centre Jean-Bérard, Collège de France, 2003, pp. 145–81.

10 Translated by Frances Muecke from *Servii Grammatici qvi fervntvr in Vergilii carmina commentarii: Aeneidos librorvm VI–XII commentarii*, recensvit G. Thilo et al., Lipsiae: B. G. Teubneri, 1883–4.

11 Antonio Di Lisio, Filippo Russo and Michele Sisto, 'Un itinéraire entre géotourisme et sacralité en Irpinie (Campanie, Italie)', *Physio-Géo*, 4, 2010, pp. 129–49.

12 For example, Philipp Clüver (Cluverius), *Italia antiqua*, Elzevir, 1624, pp. 1086–7.

13 Corinna Guerra, 'If you don't have a good laboratory, find a good volcano: Mount Vesuvius as a natural chemical laboratory in eighteenth-century Italy', *Ambix*, 62/3, 2015, pp. 245–65.

14 Eugenia Shanklin, 'Beautiful deadly lake Nyos: The explosion and its aftermath', *Anthropology Today*, 4/1, 1988, pp. 12–14.

15 Bole Butake, *Lake God and Other Plays*, Yaounde: Editions CLE, 1999. See also Eugenia Shanklin, 'Exploding lakes in myths and reality: An African case study', in Luigi Piccardi and Bruce W. Masse, eds, *Myth and Geology*, London: The Geological Society Editions, 2007, pp. 165–76.

16 Augustin Berque, *Les Raisons du paysage: De la Chine antique aux environnements de synthèse*, Paris: Hazan, 1995.

17 Luigi Piccardi et al., 'Scent of a myth: Tectonics, geochemistry and geomythology at Delphi (Greece)', *Journal of the Geological Society*, 165, 2008, pp. 5–18.

18 From Charon, the ferryman of the underworld, an old man who, for a fee, carries the wandering shadows of the dead in his boat across the river Acheron to Hades.

19 Pliny, *Natural History*, II, 95, trans. H. Rackham, Cambridge, MA: Harvard University Press, 1983, Loeb Classical Library eBooks Collection. See also Servius, *Servii Grammatici*, VII, 563–71, which mentions a list of geographical locations of the gates of Hades drawn up by Varron.

20 Cicero, *On Divination*, I, 36, trans. William Armistead Falconer, Cambridge, MA: Harvard University Press, 2014, Loeb Classical Library eBooks Collection, p. 311.

21 Virgil, *The Aeneid*, p. 134.

22 Ibid.

23 '[P]rophetic caves, those intoxicated by whose exhalations foretell the future, as at the very famous oracle at Delphi. In these matters what other explanation could any mortal man adduce save that they are caused by the divine power of that nature which is diffused throughout the universe, repeatedly bursting out in different ways?' Pliny, *Natural History*, II, 95.

24 'I believe, too, that there were certain subterranean vapours which had the effect of inspiring persons to utter oracles. ... It is in this state of exaltation that many predictions have been made ... and the cryptic utterances of Apollo were expressed in the same form.' Cicero, *On Divination*, I, 50.

25 Piccardi et al., 'Scent of a myth'; Jelle Zeilinga de Boer, 'The Oracle at Delphi: The Pythia and the pneuma. Intoxicating gas finds, and hypotheses', in Philip Wexler, ed., *History of Toxicology and Environmental Health*, Vol. 1, *Toxicology in Antiquity*, Amsterdam: Academic Press, 2014, pp. 83–9; Hardy Pfaz et al., 'The ancient gates to hell and their relevance to geogenic CO_2', ibid., pp. 92–116.

26 'Cave' (*grotte*), entry in *Encyclopédie ou Dictionnaire raisonné des sciences, des arts et des métiers*, Vol. VII, 1751, pp. 967–9, attributed to Baron d'Holbach and Louis de Jaucourt. In addition, there are the no less incredible corrections made by Abbé Jean Saas, canon at Rouen cathedral, in his *Lettres sur l'Encyclopédie to serve as a supplement to the seven volumes of this dictionary* (Isaac Tirion, 1764). He comments: 'The encyclopedists give the Dog Cave the name *Fameuse Mofeta*. This is apparently how they translate the Latin word *Mephitis*. If they ever give this Article *Mophete*, it will be a curious Article. They provide a sample here' (p. 154). The article 'moufettes' appears in a later volume of the *Encyclopedia* (Vol. X, pp. 778–80). It is attributed to Baron d'Holbach, who specified how many different types of *moufettes* there are to be distinguished: sulphurous *moufettes* (which, according to him, is what the Dog Cave is), arsenical or mercurial *moufettes* and carbonaceous *moufettes*. [The University of Michigan's online Collaborative Translation Project has not yet reached 'grotte' – Trans.]

27 Seneca, *Natural Questions*, VI, 23, trans. Thomas H. Corcoran, Cambridge, MA: Harvard University Press, 1972, Loeb Classical Library 457, p. 193.

28 'There is no reason you should think this happened to those sheep because of fear. For they say that a plague usually occurs after a great earthquake, and this is not surprising. For many death-carrying elements lie hidden in the depths.

The very atmosphere there, which is stagnant either from some flaw in the earth or from inactivity and the eternal darkness, is harmful to those breathing it. Or, when it has been tainted by the poison of the internal fires and is sent out from its long stay it stains and pollutes this pure, clear atmosphere and offers new types of disease to those who breathe the unfamiliar air. ... I am not surprised that sheep have been infected – sheep which have a delicate constitution – the closer they carried their heads to the ground, since they received the afflatus of the tainted air near the ground itself. If the air had come out in greater quantity it would have harmed people too; but the abundance of pure air extinguished it before it rose high enough to be breathed by people.' Seneca, *Natural Questions*, VI, pp. 205–7.

29 Seneca, *Natural Questions*, VI, 23, pp. 207–8.

30 Servius recalls that the name *Amsanctus* means 'inviolable': 'of a place "amsanctus", i.e. inviolable on all sides [*id est omni parte sancti*]: he [Virgil] says that it is surrounded by forests and a resounding stream'. Ibid., VII, p. 565.

31 See Lejeune, 'Méfitis, déesse osque'.

32 Filippo Coarelli, ed., *Pompeii*, trans. Patricia A. Cockram, New York: Barnes & Noble Books, 2006.

Chapter 2 An indescribable air

1 Even today, in France [and elsewhere – Trans.], chemistry textbooks continue to present this principle as 'Lavoisier's law', implying firstly that it is a law rather than a principle, and secondly that Lavoisier is the author. Yet this principle had been accepted as fundamental since the ancient physicists ('nothing is born of nothing and nothing returns to nothing' was first forged in Lucretius' *De rerum natura*). And Van Helmont, well before Lavoisier, used it to decipher what happens in chemical reactions in his *Ortus medicinae* (Amsterdam: Elsevier, 1648; reprinted in Brussels: Éditions Culture et civilisation, 1966).

2 William Newman and Lawrence Principe, *Alchemy Tried in the Fire: Starkey, Boyle and the Fate of Helmontian Chymistry*, Chicago: University of Chicago Press, 2002, pp. 58–91 and pp. 273–4.

3 Jan Baptist Van Helmont cited by Boghurst in Paulo Alves Porto, 'Os primeros desinvolvimientos de conceito helmontiano de gàs', *Quimica Nova*, 26, 2003; dx.doi.org/10.1590/S0100-40422003000100025.

4 See Bernadette Bensaude-Vincent and Isabelle Stengers, *Histoire de la chimie*, Paris: La Découverte, 1992, p. 33. Gérard Borvon in his *Histoire du carbone et du CO$_2$* (Paris: Vuibert, 2013) speaks of a 'race for airs'.

5 This reaction had been known for a long time and is even the origin of the term

'calcination' (transformation into lime) and the word 'limestone'. Quicklime reacts strongly with water and gives a slaked lime (calcium hydroxide $Ca(OH)_2$), which when dissolved in water becomes limewater. If we expose this caustic soda to caustic potash (potassium hydroxide), we give rise to fixed air and the limestone comes out as a precipitate.

6 Joseph Priestley, *Directions for Impregnating Water*, Second Edition, London: J. Johnson, 1775 (https://www.gutenberg.org/cache/epub/29734/pg29734 -images.html).

7 Ibid., p. 52.

8 Sir John Pringle, *Six Discourses, Delivered by Sir John Pringle, Bart. When President of the Royal Society; On Occasion of Six Annual Assignments of Sir Godfrey Copley's Medal*, London: W. Strand and T. Caddell, 1783, pp. 34–5 (https:// archive.org/details/b28758018/page/n5/mode/2up?ref=ol&view=theater).

9 Arthur Donovan, ed., *The Chemical Revolution: Essays in Reinterpretation*, *Osiris*, Second Series, Vol. 4, Philadelphia: History of Science Society, University of Pennsylvania, 1988; Ferdinando Abbri, 'The chemical revolution: A critical reassessment', *Nuncius*, 4/2, 1989, pp. 303–15; Bernadette Bensaude-Vincent, *Lavoisier: Mémoires d'une révolution*, Paris: Flammarion, 1993.

10 Antoine-Laurent Lavoisier, *Essays Physical and Chemical*, trans. Thomas Henry, London: J. Johnson, 1776, p. 198.

11 Antoine Lavoisier, mémoire 'Sur l'affinité du principe oxygine', 20 December 1783, in *Œuvres*, Vol. II, Paris: Imprimerie impériale, 1862, p. 546.

Chapter 3 Between diamond and coal

1 Académie des sciences, 'Plumitif Lavoisier', 27 July 1771. For this entire section we are drawing on the work of Christine Lehmann: 'What is the "true" nature of diamond?', *Nuncius*, 31/2, 2016, pp. 361–407; Christine Lehmann, 'À la recherche de la nature du diamant: Guyton de Morveau successeur de Macquer et Lavoisier', *Annales historiques de la Révolution française*, 1, 2016, pp. 81–107.

2 'Pierres fines' entry in Diderot and D'Alembert's *Encyclopédie*.

3 Minutes, Académie des sciences, dossier from 29 April 1772.

4 Ibid.

5 'Sur la destruction du diamant au grand verre brûlant de Tschirnhaus, connu sous le nom de Lentille du Palais royal', *Mémoires de l'Académie royale des sciences* (1772), 1776, II, pp. 591–616. Macquer gives another account of these experiments in the article 'Diamant' in his *Dictionnaire de chymie*, Vol. I, 1778.

6 Louis-Bernard Guyton de Morveau, *Annales de chimie*, 32, 30 vendémiaire an VIII (22 October 1799), pp. 208–11. See also the article 'Diamant' in *Dictionnaire de chymie de l'Encyclopédie méthodique*, Vol. IV, an VIII, 1805,

pp. 152–7. [With this reference, the authors have retained the French post-revolutionary dating system 'vendémiaire an VIII' and 'an VIII', as well as the conventional year – Trans.]

7   Louis-Bernard Guyton de Morveau, 'Air' entry in *Dictionnaire de chymie, de pharmacie et de métallurgie de l'Encyclopédie méthodique*, Vol. I, 1786, p. 741.

8   Louis-Bernard Guyton de Morveau, 'Mémoire sur les dénominations chymiques, la nécessité d'en perfectionner le système et les règles pour y parvenir', *Observations sur la physique, sur l'histoire naturelle et sur les arts*, 19, May 1782, pp. 370–82. Also published as a pamphlet in Dijon in 1782.

9   Louis-Bernard Guyton de Morveau et al., *Méthode de nomenclature* (1787); translated as *A Translation of the Table of Chemical Nomenclature*, Second Edition, London: Cooper and Wilson, 1799.

10 Entry for 'Carbon' in the *Dictionnaire de chymie de l'Encyclopédie méthodique*, Vol. III, 1803, p. 47.

11 Ibid., p. 50.

12 Ibid., p. 48.

13 Ibid., p. 51.

14 Marc-Antoine Eidous, 'Charbon' entry in *Encyclopédie ou Dictionnaire raisonné des sciences et des techniques*, Vol. III, 1752 (https://xn--encyclopdie-ibb.eu/index.php/science/950782828-mecanique/773546137-CHARBON).

15 Ibid.

16 Smithson Tennant, 'On the nature of diamond', *Philosophical Transactions of the Royal Society*, 87, 1797, pp. 123–7.

17 Guyton de Morveau et al., *Méthode de nomenclature*, pp. 81 and 84.

18 Ibid., p. 84.

19 Guyton de Morveau, 'Diamant' entry in *Dictionnaire de chimie*.

20 Hélène Metzger, *La Genèse de la science des cristaux*, Second Edition, Paris: Librairie scientifique et technique Albert Blanchard, 1969 [1918]; Bernard Maitte, *Histoire des cristaux*, Paris: ADAPT-SNES, 2014.

21 Jacob Berzelius, 'Anorganische Chemie: Isomerie', *Jahre-Bericht*, 20, 1841, pp. 7–13. See also William B. Jensen, 'The origin of the term "allotrope"', *Journal of Chemical Education*, 83/6, 2006, pp. 838–9.

22 Jacob Berzelius, *Traité de chimie*, trans. A.-J.-L. Jourdan, Paris: Firmin Didot Frères, 1856, pp. 21–2.

23 Today, allotropy is defined as a form of polymorphism that concerns substances formed from a single element.

24 Alfred Naquet, *De l'allotropie et de l'isomérie*, Paris: J.-B. Baillière, 1860.

25 Ibid., pp. 62–5.

26 Ibid., p. 60.

Chapter 4 An exemplary element

1 Gustave Flaubert, *Bouvard and Pecuchet*, trans. Mark Polizzotti, Illinois: Dalkney Archive Press, 2005, p. 52; see also Mitsumasa Wada, 'L'épisode de la chimie dans *Bouvard et Pécuchet* de Flaubert, ou comment narrativiser une ambiguïté scientifique', *Item* (http://www.item.ens.fr/articles-en-ligne/lepisode-de-la-chimie-dans-bouvard-et-pecuchet-de-flaubert-o/).

2 Bernadette Bensaude-Vincent, Antonio Garcia Belmar and José-Ramon Bertomeu, *L'Émergence d'une science des manuels: Les livres de chimie en France (1789–1852)*, Paris: Éditions des archives contemporaines, 2003.

3 Bensaude-Vincent and Stengers, *Histoire de la chimie*, pp. 125–39. Christopher Hamlin, 'The city as a chemical system? The chemist as urban environmental professional in France and Britain, 1780–1880', *Journal of Urban History*, 33, 2007, pp. 702–28.

4 Roald Hoffmann, *The Same and Not the Same*, New York: Columbia University Press, 1995, p. xvi.

5 Thomas Kuhn, *The Structure of Scientific Revolutions*, Chicago: University of Chicago Press, 1962, postface.

6 Victor Regnault, *Cours élémentaire*, Paris: Langlois, Vol. I, 1840, p. 3.

7 Antoine Lavoisier, *Traité élémentaire de chimie*, Paris: Cuchet, 1789, p. xvii.

8 At the beginning of the nineteenth century, the category of imponderable bodies (light, heat, electric fluid and magnetic fluid) was added. See Bensaude-Vincent et al., *L'Émergence d'une science des manuels*.

9 Regnault admitted that a 'natural classification' (taking into account all properties) would be philosophically more satisfying than this artificial classification (based on a single property). He even proposed a hybrid system of classification that combined a version he had revised and corrected of the artificial classification of metals proposed by Louis Jacques Thenard with the natural classification of non-metals proposed by Jean-Baptiste Dumas.

10 On the controversy between natural and artificial classification among textbook authors in the nineteenth century, see Bernadette Bensaude-Vincent, Antonio Garcia and José R. Bertomeu, 'Looking for an order of things: Textbooks and chemical classifications in nineteenth-century France', *Ambix*, 49, 2002, pp. 227–50.

11 See Stephanie Dord-Crouslé, *Bouvard et Pécuchet de Flaubert: Une encyclopédie critique en farce*, Paris: Belin, 2000.

12 Dimitri Mendeleev, *Principles of Chemistry*, New York: Collier, 1901.

13 Ibid., p. 22.

14 Ibid., p. 23.

15 Ibid., p. 24.

16 Dimitri Mendeleev, 'The relation between the properties of the atomic weight of the elements' (1871), in H. M. Leicester and H. S. Klickstein, eds, *Sourcebook in Chemistry 1400–1900*, New York: Dover, 1953, p. 693.

17 This law, stated simultaneously by Avogadro and Ampere in 1811, specifies that equal volumes of different gases contain the same number of molecules.

18 Mendeleev, 'The relation between the properties of the atomic weight of the elements', p. 493.

19 On this controversy see Eric R. Scerri, 'Correspondence and reduction in chemistry', in Steven French and Harmke Kamminga, eds, *Correspondence, Invariance and Heuristics*, Dordrecht: Kluwer Academic Publishers, 1993, pp. 45–64.

20 Georges Urbain, *Les Notions fondamentales d'élément chimique et d'atome*, Paris: Gauthier-Villars, 1925, p. 9.

21 Mendeleev, *Principles of Chemistry*, p. 2.

22 On the concept of 'metachemistry', see Gaston Bachelard, *The Philosophy of No: A Philosophy of the New Scientific Mind*, trans. G. C. Waterston, London: Orion Press, 1968 [1940].

23 Friedrich Paneth, 'The epistemological status of the concept of element' (1931), *British Journal for the Philosophy of Science*, 13, 1962, pp. 1–14 and 144–160, reprinted in *Foundations of Chemistry*, 5, 2003, pp. 113–45. Cf. Klaus Ruthenberg, 'Friedrich Adolph Paneth (1887–1938)', *Hyle, An International Journal in the Philosophy of Chemistry*, 3, 1993, pp. 103–6.

24 Friedrich Paneth, in Herbert Dingle, G. R. Martine and Eva Paneth, eds, *Chemistry and Beyond: A Selection of the Writings of the Late F. Paneth*, New York: Interscience Publishers, 1964, pp. 66–7.

25 For the allusion to Newton and Kant, see Dimitri Mendeleev, 'The periodic law of the chemical elements' (1871), in *Mendeleev on the Periodic Law: Selected Writings, 1869–1905*, ed. William B. Jensen, Mineola, NY: Dover, 2002, p. 159.

Chapter 5 Carbon liberates itself

1 Gaston Bachelard, *Le Pluralisme cohérent de la chimie moderne*, Paris: Vrin, 1973 [1930].

2 François Leprieur, 'Les conditions de la constitution d'une discipline scientifique: la chimie organique en France (1830–1880)', Ph.D. dissertation at the Université de Paris 1, 1977.

3 Alan J. Rocke, *The Quiet Revolution: Hermann Kolbe and the Science of Organic Chemistry*, Berkeley: University of California Press, 1983.

4 Jean Jacques, 'La thèse de doctorat d'Auguste Laurent et la théorie des

combinaisons organiques (1836)', *Bulletin de la Société chimique de France*, 1954, pp. 31–9; Jean Jacques, *Marcellin Berthelot: Autopsie d'un mythe*, Paris: Belin, 1987.

5 Alan J. Rocke, *Nationalizing Chemistry and the Battle for French Chemistry*, Cambridge, MA: MIT Press, 2001.

6 Ursula Klein, *Experiments, Models, Paper Tools: Cultures of Organic Chemistry in the Nineteenth Century*, Stanford: Stanford University Press, 2003. Sacha Tomic, *Aux origines de la chimie organique: Méthodes et pratiques des pharmaciens et des chimistes (1785–1835)*, Rennes: Presses Universitaires de Rennes, 2010.

7 For this section we have drawn heavily on Tomic, *Aux origines de la chimie organique*.

8 Klein, *Experiments, Models, Paper Tools*, p. 222.

9 Antoine Fourcroy, *Système des connaissances chimiques et de leurs applications aux phénomènes de la nature et de l'art*, Paris: Baudouin, 1801 (11 vols).

10 Louis Jacques Thenard, *Traité de chimie élémentaire, théorique et pratique*, Paris: Crochard [1813–16 (4 vols)], Sixth Edition, 1835 (5 vols); see Bensaude-Vincent et al., *L'Émergence d'une science des manuels*.

11 Ferdinand Hoefer's expression in *Nomenclature et classifications chimiques*, Paris: Baillière, 1845.

12 Thenard divided the thirty-eight known metals into six classes according to whether or not they absorbed oxygen, at high temperatures or not, and whether or not they decomposed water. Oxygen thus provides an elegant means of ordering knowledge of inorganic chemistry. It is less effective in the field of plant and animal chemistry. Thenard adopted a classification based on the chemical composition of the immediate principles, according to the respective proportions of hydrogen and oxygen, i.e. three classes: (1) more oxygen than in water; (2) the same proportion; (3) less oxygen.

13 Tomic, *Aux origines de la chimie organique*, p. 243.

14 Auguste Comte, *The Positive Philosophy of August Comte*, trans. Harriet Martineau, London: George Bell, 1896 [1835], Ch. V.

15 Ursula Klein, ed., *Tools and Modes of Representation in the Laboratory Sciences*, Dordrecht: Kluwer, 2001.

16 Alan J. Rocke, 'Organic analysis in a comparative perspective: Liebig, Dumas and Berzelius (1811–1837)', in Frederic L. Holmes and Trevor Levere, eds, *Instruments and Experiments in the History of Chemistry*, Cambridge, MA: MIT Press, 2000, pp. 273–310.

17 Unlike previous instruments that burned small samples of compounds and then collected the carbonic acid as a gas to measure its volume, the *kaliapparat* captures the carbonic acid resulting from combustion in the condensed phase

for gravimetric measurement. The trick is to use a potassium hydroxide solution that strongly attracts the carbonic acid and condenses it into potassium carbonate. By weighing the apparatus before and after the analysis, the increase in weight is attributed to the CO_2 released by the sample. The volume of nitrogen is then measured. This apparatus has proven to be very effective in determining the formula of alkaloids.

18 Auguste Laurent, *Méthode de chimie*, Paris: Firmin Didot, 1854, p. 1.

19 Marcellin Berthelot, *La Synthèse chimique*, Paris: Librairie Germer Baillière, 1876, pp. 215–17.

20 Ibid., pp. 10–11.

21 Ibid., pp. 11–12.

22 'Among the simple bodies that have an effect on oils, I will mention chlorine. When it is applied in a gaseous state on oils, it removes from them a portion of hydrogen with which it combines to form hydrochloric acid, which can be collected; and at the same time, part of the chlorine combines with the oil and takes the place of the hydrogen removed. . . . Wax is bleached with chlorine; but it combines with the wax, and by burning it, it spreads thick hydrochloric acid vapours in the rooms. This means of bleaching must be abandoned.' Joseph Gay-Lussac, *Cours de chimie*, 1828, twenty-eighth lesson, pp. 11 and 22, quoted in French in John Partington, *A History of Chemistry*, London: Macmillan, 1964, p. 360.

23 Jean-Baptiste Dumas, 'Considérations générales sur la composition théorique des matières organiques', *Journal de pharmacie*, 5, 20 May 1834, pp. 261–94, p. 269.

24 Ibid., p. 268.

25 Ibid., p. 289.

26 Ibid., p. 294.

27 Christiane Buès, 'Histoire du concept de mole: À la croisée des disciplines physique et chimie', Lille: Atelier national de reproduction des thèses, 2001.

Chapter 6 A relational being

1 Charles-Adolphe Wurtz, *The Atomic Theory*, trans. Edward Cleminshaw, New York: D. Appleton & Co., 1881, pp. 196–223.

2 Ibid., pp. 215–16. Translation modified.

3 Klein, *Experiments, Models, Paper Tools*.

4 Berthelot, *La Synthèse chimique*, p. 167.

5 Charles Gerhardt, *Traité de chimie organique*, Vol. IV, Paris: Firmin Didot, 1854–56, p. 566.

6 Wurtz, *The Atomic Theory*, p. 313.

7 At least those chemists who accepted Avogadro's hypothesis that equal volumes

of different gases contain the same number of molecules under the same conditions of temperature and pressure.

8 Charles-Adolphe Wurtz, *La Théorie atomique*, Paris: Librairie Germer Baillière, 1879, p. 158. Passage not available in the English edition.

9 Alan J. Rocke, *Image and Reality: Kekulé, Kopp, and the Scientific Imagination*, Chicago: University of Chicago Press, 2010.

10 See van't Hoff's dedication to Le Bel in *Chemistry in Space*, trans. and ed. J. E. Marsh, Oxford: Clarendon Press, 1981 [1887].

11 Jean-Claude Compain, 'Les travaux de Le Bel et van't Hoff de 1874 et notre enseignement', *Bulletin de l'Union des physiciens*, 741, 1992, pp. 295–311.

12 For example, in the chlorobromoethane molecule ($CH_3CHBrCl$) resulting from the substitution of a bromine atom and then a chlorine atom for two hydrogen atoms, there is one carbon atom bonded to three hydrogen atoms and one carbon atom bonded to one hydrogen atom, one bromine atom and one chlorine atom.

13 Achille Le Bel, 'Sur les relations qui existent entre les formules atomiques des corps organiques et le pouvoir rotatoire de leurs dissolutions', *Bulletin de la Société chimique*, 1874, p. 337, cited by van't Hoff, *Chemistry in Space*, pp. 2–3.

14 Achille Le Bel, 'Les relations du pouvoir rotatoire avec la structure moléculaire', *Revue scientifique*, 48/20, 14 November 1891, p. 612.

15 Peter J. Ramberg and Geert J. Somsen, 'The young J. H. van't Hoff: The background to the publication of his 1874 pamphlet on the tetrahedral carbon atom, together with a new English translation', *Annals of Science*, 58, 2001, pp. 51–74.

16 Jacobus van't Hoff, 'Essai d'un système de formules atomiques à trois dimensions et la relation entre le pouvoir rotatoire et la constitution chimique qui en découle' (1874), in *Chemistry in Space*, p. 9.

17 Hermann Kolbe, 'Zeichen der Zeit II', cited and translated by van't Hoff, *Chemistry in Space*, p. 17.

18 H. A. M. Snelders, 'The reception of J. H. van't Hoff's theory of the asymmetric carbon atom', *Journal of Chemical Education*, 51, 1974, pp. 2–7.

19 Louis Pasteur, 'La dissymétrie moléculaire', lecture on 22 December 1883, in Jean Jacques, *Sur la dissymétrie moléculaire: L. Pasteur, J. H. van't Hoff, L. Werner*, Paris: Christian Bourgeois, 1986, p. 98.

20 Ibid., p. 103.

21 Frederic S. Kipping, 'Organic derivatives of silicon', *Proceedings of the Royal Society A*, 159, 1937, pp. 139–48. William B. Jensen and Peter J. T. Morris, 'From chemical theory to industrial chemistry: The eclectic career of Geoffrey Martin', *Bulletin of the History of Chemistry*, 41, 2016, pp. 19–37.

22 Silane, for example, is the siliceous structural analogue of methane. Like carbon, silicon is tetravalent. In particular, Kipping identified the family of 'silicones' with the empirical formula R_2SiO on the model of the ketones R_2CO.

23 The hypothesis was based on the idea that organic carbon compounds are only stable in one temperature and pressure range, whereas in a higher temperature and pressure range, prior to our geological era, silicon could have given rise, together with phosphorus and sulphur, to a protoplasm now petrified in the rocks that make up the earth.

24 Alfred Werner, 'Sur les composés métalliques à dissymétrie moléculaire', lecture to the Société chimique de France, 24 May 1912, in Jacques, *Sur la dissymétrie moléculaire*, pp. 223–54.

25 Nuno Manuel Catanheiro de Figueiredo, 'Under the carbon spell: Diborane's puzzling structure and the emergence of boron', Master's thesis in the history and philosophy of the sciences, University of Lisbon, 2011.

26 This is the medium that causes the carbon atom to go from the ground state to the excited state by absorption of a photon. It becomes quadrivalent when it changes from the $2s^2 2p^2$ to the $2s^1 2p^3$ electronic configuration.

27 Gaston Bachelard, *Le Matérialisme rationnel*, Paris: PUF, 1950, p. 148.

28 Ian Hacking, *Representing and Intervening: Introductory Topics in the Philosophy of Natural Science*, Cambridge: Cambridge University Press, 1983.

29 Nancy Cartwright, 'Where do the laws of nature come from?', *Dialectica*, 51, 1997, pp. 56–78.

30 According to Gibson, the animal in its environment does not perceive forms of objects and relations between these forms, but possible actions related to the objects. We – animals in general – do not directly perceive the properties of the things around us but rather their meaning, their valence, the interest they may have for us in a specific situation.

31 James J. Gibson, *The Ecological Approach to Visual Perception*, New York: Psychology Press Classic Edition, 2015 [1986], p. 130.

32 Ludwig F. Haber, *The Chemical Industry during the Nineteenth Century*, Oxford: Oxford University Press, 1958.

33 Fred Aftalion, *A History of the International Chemical Industry, From the Early Days to 2000*, Philadelphia: Chemical Heritage Foundation, 2005.

34 Hermann Staudinger, 'Die Chemie der hochmolekularen organischen Stoffe im Sinne der Kékuleschen Strukturlehre', *Berichte der Deutschen Chemischen Gesellschaft*, 59, 1926, pp. 3019–43, p. 3043.

35 Yasu Furukawa, 'Polymer chemistry', in John Krige and Dominique Pestre, eds, *Science in the Twentieth Century*, Amsterdam: Harwood Academic Publishers, 1997, pp. 547–63.

36 See Ch. 12.

37 See Yves Chauvin, 'Pourquoi la métathèse a remporté le Nobel', interview with Franck Daninos, *La Recherche*, 394, 2006, p. 61.

Chapter 7 Welcome to the nanoworld

1 'There's a joke amongst nanotech insiders that the most accurate definition of "nano" is: "a tiny manufactured prefix engineered into funding proposals to exploit the unusually generous properties of science funds occurring at the nano-scale."' ETC Group, *NanoGeoPolitics*, ETC Group Special Report (Communiqué 89), July/August 2005, p. 31.

2 Bernadette Bensaude-Vincent, *Les Vertiges de la technoscience*, Paris: La Découverte, 2009.

3 Hugh Colquhoun, 'Notice of a new Form of Carbon supposed to be the pure Metallic Basis of the Substance; and also of several other interesting Aggregations of Carbon, especially in so far as they elucidate the History of certain Carbonaceous Products found in Coal Gas Manufactories', *Annals of Philosophy*, 12/28, 1826, pp. 1–13, pp. 1–2.

4 English Patent No. 5173 (1825).

5 Colquhoun, 'Notice of a new Form of Carbon', p. 3.

6 Plumbago is the old name for almost pure graphite, which is mostly used to make pencil leads. See Ch. 9.

7 Paul and Léon Schützenberger, 'Sur quelques faits relatifs à l'histoire du carbone', *Comptes rendus hebdomadaires de l'Académie des sciences*, 111, 1890, pp. 774–8.

8 Retort carbon, or gas carbon, is a very pure and dense variety of amorphous carbon obtained by gaseous distillation of bitumen or during the manufacture of lighting gas from coal. Renowned for its electrical properties, it was used as an electrode in 1841 by Robert W. Bunsen in a 'carbon battery'.

9 Schützenberger and Schützenberger, 'Sur quelques faits relatifs à l'histoire du carbone', p. 777.

10 This is, at least, the image of the École municipale de physique et de chimie as portrayed in Jean-Noël Fenwick's famous play (later a film) about the discovery of radium, *Les Palmes de monsieur Schutz*.

11 Constant and Henri Pélabon, 'Sur une variété de carbone filamenteux. Note de MM. Constant et Henri Pélabon, présentée par M. Moissan', *Comptes rendus hebdomadaires de l'Académie des sciences*, 137, 1903, pp. 706–8.

12 Brigitte Schroeder-Gudehus and Anne Rasmussen, eds, *Les Fastes du progrès: Le guide des expositions universelles 1851–1992*, Paris: Flammarion, 1997.

13 Robert Friedel and Paul Israel, *Edison's Electric Light: Biography of an Invention*, New Brunswick: Rutgers University Press, 1986, pp. 115–17.

14 Flash carbonization consists of carbonizing the fibre by means of the electric arc used in early electric lighting.

15 Jinquan Wei et al., 'Carbon nanotube filaments in household light bulbs', *Applied Physics Letters*, 84/14, 2004, pp. 4869–71.

16 W. R. Davis, R. J. Slawson and G. R. Rigby, 'An unusual form of carbon', *Nature*, 171, 1953, p. 756.

17 L. Radushkevich and V. M. Lukyanovich, 'The carbon structure formed by thermal decomposition of carbon monoxide in contact with iron', *Zhurnal Fizicheskoi Khimi*, 26, 1952, pp. 88–95 (in Russian).

18 See Ch. 8.

19 Mats Hillert and Nils Lange, 'The structure of graphite filaments', *Zeitschrift für Kristallographie – Crystalline Materials*, 111/1-6, 1958, pp. 24–34.

20 Marijana Moni, 'Preparation and physical characterization of carbon nanotubes-SU8 composites', Thesis No. 5248, École polytechnique de Lausanne, 2011.

21 Marc Monthioux and V. L. Kuznetsov, 'Who should be given the credit for the discovery of carbon nanotubes?', *Carbon*, 44, 2006, pp. 1621–5.

22 Roger Bacon, 'Production and properties of graphite whiskers', *Bulletin of the American Physics Society*, 2, 1957, p. 131.

23 Roger Bacon, 'Growth, structure, and properties of graphite whiskers', *Journal of Applied Physics*, 31, 1960, pp. 283–90.

24 American Chemical Society, 'High performance carbon fibers', *National Historic Chemical Landmarks*, 2003 (http://www.acs.org/content/acs/en/education/whatischemistry/landmarks/carbonfibers.html).

25 On the age of plastics and the composite technology that has expanded the plastics market, see Ch. 12.

26 Morinobu Endo, Agnès Oberlin and Tsuneo Koyama, 'Filamentous growth of carbon through benzene decomposition', *Journal of Crystal Growth*, 32, 1976, pp. 335–49, p. 335.

27 Morinobu Endo, 'Grow carbon fibers in the vapor phase', *Chemical Technology*, 18, 1988, pp. 568–76.

28 Based on a patent filed in 1982 for the production process developed with Oberlin (Morinobu Endo and Tsuneo Koyama, 'Preparation of carbon fibre by vapor phase method', Patent JPS58180615A, 1982), the applications concern electric batteries, for which Endo's fibres are mass-produced and intercalated in the anode of lithium ion batteries.

29 Harold W. Kroto et al., 'The detection of HC_9N in interstellar space', *The Astrophysical Journal*, 223, 1978, pp. 105–7.

30 Cyrus C. M. Mody, 'The diverse ecology of electronic materials', *Cahiers François Viète*, III/2, 2017, pp. 217–41, p. 232.

31 Harold Kroto et al., 'C$_{60}$: buckminsterfullerene', *Nature*, 318, 1985, pp. 162–3.

32 Kroto, as cited by Jeffrey I. Seeman and Stuart Cantrill, 'Wrong but seminal', *Nature Chemistry*, 8, March 2016, pp. 193–200.

33 David E. H. Jones, 'Hollow molecules', *New Scientist*, 32, 1966, p. 245.

34 Eiji Osawa, 'Superaromaticity', *Kagaku*, 25, 1970, pp. 854–63 (in Japanese).

35 D. A. Bochvar and E. G. Gal'pern, 'On hypothetical systems: Carbododecahedron, s-icosahedron and carbo-s-icosahedron', *Doklaldy Akademii Nauk SSSR*, 209/3, 1973, pp. 610–12 (in Russian); Robert A. Davidson, 'Spectral analysis of graphs by cyclic automorphism subgroups', *Theoretica Chimica Acta*, 58, 1981, pp. 193–5.

36 Tony A. D. J. Haymet, 'Footballene: A theoretical prediction for the stable, truncated icosahedral molecule C$_{60}$', *Chemical Physics Letter*, 122, 1985, pp. 421–4.

37 Orville Chapman, quoted by Cyrus C. M. Mody, '1990: Buckyball & carbon nanotubes', in Raymond J. Giguere, ed., *Molecules That Matter*, Philadelphia: Chemical Heritage Foundation, 2008, pp. 159–76, p. 166.

38 Peter J. F. Harris, *Carbon Nanotubes and Related Structures: New Materials for the Twenty-First Century*, Cambridge: Cambridge University Press, 1999, p. 2.

39 Wolfgang Krätschmer et al., 'Solid C$_{60}$: A new form of carbon', *Nature*, 347, 1990, pp. 354–8.

40 Sumio Iijima, 'Helical microtubules of graphitic carbon', *Nature*, 354, 1991, pp. 56–8.

41 Even those who seek to relativize the discovery by invoking earlier observations of tubes in the carbon filaments described above cite Iijima's paper and paradoxically increase the number of citations.

42 The term 'nanotube' was coined as short form for 'hollow graphitic tubules of nanometre dimensions' by the Franco-Norwegian physical chemist Thomas Ebbesen, working at the same time as Iijima in the same research institution (the NEC Corporation's Tsukuba Fundamental Research Laboratories) in a 1992 paper reporting 'large-scale' synthesis (gram quantities) of these objects (Thomas W. Ebbesen and P. M. Ajayan, 'Large-scale synthesis of carbon nanotubes', *Nature*, 358, 1992, pp. 220–2). Iijima only began using it in 1993 (Sumio Iijima and Toshinari Ishahashi, 'Single-shell carbon nanotubes of 1-nm diameter', *Nature*, 363, 1993, pp. 603–5).

43 Interview of Morinobu Endo by Bernadette Bensaude-Vincent, *Sciences, Histoire orale* (https://www.sho.espci.fr/spip.php?article48).

44 Sumio Iijima, 'Direct observation of the tetrahedral bonding in graphitized carbon black by high-resolution electron microscopy', *Journal of Crystal Growth*, 50, 1980, pp. 675–83; 'High resolution electron microscopy of some carbonaceous materials', *Journal of Microscopy*, 119, 1980, pp. 99–111.

45 Sacha Loeve, 'Nanocarbons', in Bensaude-Vincent, ed., *Between Nature and Society*, pp. 129–42.

46 The geometric language of description devised by Iijima had soon after been extended into a *design language* to obtain tubes having the desired electronic properties by Noriaki Hamada. Noriaki Hamada et al., 'New one-dimensional conductors: Graphitic microtubules', *Physical Review Letters*, 68/10, 1992, pp. 1579–81. On Hamada's notation, see Sacha Loeve, 'Point and plane to line: The ontography of carbon materials', *Cahiers François Viète*, III/2, 2017, pp. 183–216, pp. 203–4.

Chapter 8 Strategic materials

1 https://graphene-flagship.eu/.

2 See Ch. 9.

3 P. L. Walker, 'Carbon. And old but new material', *Scientific American*, 50/2, 1962, pp. 259–93.

4 Robert L. Carter and Richard Eggleton, 'Moderator graphite for high temperature reactor', Nuclear Engineering and Manufacturing, North American Aviation Inc., Downey, California, 1 August 1955 (https://personal.ems.psu .edu/~radovic/1955/papers/1955_149.PDF).

5 Gabrielle Hecht, *The Radiance of France: Nuclear Power and National Identity after World War II*, Cambridge, MA: MIT Press, 2000.

6 *Reaktor Bolshoy Moshchnosti Kanalnyi* (high-power pressure tube reactor).

7 M. S. Dresselhaus and G. Dresselhaus, 'Intercalation compounds of graphite', *Advances in Physics*, 30, 1981, pp. 139–326.

8 Claire Hérold and Philippe Lagrange, 'Les réactions d'intercalation dans le graphite: Une chimie bidimensionnelle', *L'Actualité chimique*, 295–6, March–April 2006, pp. 33–7.

9 Hanns-Peter Boehm et al., 'Dunnste Kohlenstoff-Folien', *Zeitschrift für Naturforschung B*, 17, 1962, pp. 150–3; 'Surface properties of extremely thin graphite lamellae', *Proceedings of the Fifth Conference on Carbon*, London: Pergamon Press, 1962, pp. 73–80.

10 Hanns-Peter Boehm, Ralph Setton and Eberhard Stumpp, 'Nomenclature and terminology of graphite intercalation compounds', *Carbon*, 24/2, 1986, pp. 241–5; 'Nomenclature and terminology of graphite intercalation compounds – IUPAC 1994 recommendations', *Pure & Applied Chemistry*, 66/9, 1994, pp. 1893–901.

11 Rudolph Peierls, 'Quelques propriétés typiques des corps solides', *Annales de l'Institut Henri-Poincaré*, 5, 1935, pp. 177–222; L. D. Landau, 'Zur Theorei der phasenumwandlugen II', *Physikalische Zeitschrift Sowjetunion*, 11, 1937,

pp. 26–35; N. D. Mermin, 'Crystalline order in two dimensions', *Physical Review*, 176, 1968, pp. 250–4.

12 See Ch. 7.

13 That is, a thermodynamically unstable state, but kinetically stable.

14 Andre Geim, 'Nobel lecture', 2010, pp. 87–8 (https://www.nobelprize.org/prizes/physics/2010/geim/lecture/).

15 Gilles Deleuze and Félix Guattari, *Anti-Oedipus: Capitalism and Schizophrenia*, trans. Robert Hurley, Mark Seem and Helen R. Lane, Minneapolis: University of Minnesota Press, 1983 [1972].

16 Geim, 'Nobel lecture', pp. 88–9.

17 Gilbert Simondon, *On the Mode of Existence of Technical Objects*, trans. Cecile Malaspina and John Rogove, Minneapolis: Univocal/University of Minnesota Press, 2016 [1958].

18 The excited electrons create a band hole among the valence electrons; the charge carriers interact with each other and form a quasi-particle called an 'exciton' in localized states. When they de-excite, they emit light for CVS luminescent screens that absorb energies in the order of three hundred millielectronvolts.

19 According to an analysis protocol developed by the Singapore-based company 2DM to test the quality of graphene products, in most current commercial applications there is 94% graphite and 6% graphene (paper by Antonio Castro Neto, 'Graphene Week', Athens, September 2017).

20 Anonymous editorial, 'Perfectly imperfect', *Nature Nanotechnology*, 5, May 2010, p. 311.

21 Manufacturing from the bottom up, starting from the atomic and molecular building blocks.

22 Eric Drexler, *Engines of Creation: The Coming Era of Nanotechnology*, New York: Anchor Books, Doubleday, 1986.

Chapter 9 *Traces, stories and memories*

1 See Ch. 4.

2 Levi, *The Periodic Table*, chapter on 'Carbon', pp. 227–36, pp. 227, 228, 229.

3 In 1941, Levi obtained a doctorate in chemistry with a thesis on the Walden inversion (chirality inversion of a compound built around an asymmetric carbon).

4 Anne Moiroux, '*Le Système périodique* de Primo Levi: Une classification de la matière narrative', *Chroniques italiennes*, 71–2, 2003, pp. 136–47.

5 One could also speculate on the composition of the ink (carbon black?) and the paper (usually cellulose fibres $[C_6H_{10}O_5]_n$).

6 Levi, *The Periodic Table*, p. 234.

7 He was found naturally mummified (frozen and dehydrated) in the Ötztal Alps (close to the Italian Dolomites), hence his nickname 'Ötzi'. On his tattoos, see Aaron Deter-Wolf et al., 'The world's oldest tattoos', *Journal of Archaeological Science: Reports*, 5, 2016, pp. 19–24.

8 Thomas Corneille, 'Charbon', *Dictionnaire des arts et des sciences de M. D. C. de l'Académie française*, Vol. I, Paris, 1732.

9 Robert Plot, 'Some observations concerning the substance commonly called, black lead', *Philosophical Transactions of the Royal Society of London*, 20/83, 1698, p. 183.

10 Henry Petroski, *The Pencil: A History of Design and Circumstance*, New York: Knopf, 1990.

11 One of the earliest known uses of lead in writing was the ruling of parchments by ancient Egyptian scribes.

12 Alternatively, diamond can be machined at high temperature and pressure, or by laser. Ion implantation can also be used to modify its optical properties. It is used both in jewellery (where it lowers the price of diamond, which is then considered less 'natural') and in nanoelectronics to transmit information by photoemission in diamond. See Fedor Jelezko and Jörg Wrachtrup, 'Single defect centres in diamond: A review', *Physica Status Solidi (a)*, 203/13, 2006, pp. 3207–25.

13 See Ch. 8.

14 As we saw in Ch. 3.

15 Natalia Dubrovinskaia et al., 'Aggregated diamond nanorods, the densest and least compressible form of carbon', *Applied Physics Letters*, 87/8, 2005, p. 083106.

16 The diamonds act as two anvils. This device – the diamond press – is familiar to geologists because it simulates the physical conditions under the earth's crust. Most recently, it has sparked a debate about the possibility of obtaining solid metallic hydrogen. See Robert F. Service, 'Diamond vise turns hydrogen into a metal, potentially ending 80-year quest', *Science*, 355, 26 January 2017, pp. 332–3.

17 The 'system', in which the diamond is an essential link, refers to all the transduction and amplification devices.

18 The accepted half-life in 1950 was 5,568 plus or minus thirty years. It has been revised to 5,734 plus or minus forty years.

19 Greg Marlowe, 'W. F. Libby and the archeologists, 1946–1948', *Radiocarbon*, 22, 1980, pp. 1005–14; Greg Marlowe, 'Year One: Radiocarbon dating and American archeology 1947–1948', *American Antiquity*, 64, January 1999, pp. 9–32.

20 Dating at the time was based mainly on stratigraphy. Dendrochronology (dating according to the growth rings of trees) was an innovation that seems

to have been neglected by Chicago archaeologists in the 1940s. Hence their concern not to miss the boat with radiocarbon.

21 BP stands for 'before present'. The reference point for dating is the isotopic composition of the atmosphere in 1950.

22 C. P. Snow, *The Two Cultures*, Cambridge: Cambridge University Press, 1993 [1959].

23 Otherwise the sample incorporates new carbon atoms from the medium that change the $^{14}C/C_{total}$ ratio.

24 Michael Adler, *Antique Typewriters from Creed to QWERTY*, New York: Schiffer Publishing Ltd, 1997.

25 However, it is still used for signing registered mail at La Poste in France.

26 https://france-science.com/mmoire-au-graphne-les-mmoires-flash-du-futur/?print=print.

27 Katherine Bourzac, 'Electronics: Back to analogue', *Nature*, 483, 2012, pp. 34–6.

28 Andy Extance, 'How DNA could store all the world's data', *Nature*, Editorial, 537/7618, 2016 (https://www.nature.com/articles/537022a).

Chapter 10 *The resilient rise of fossil fuels*

1 Jeffrey S. Dukes, 'Burning buried sunshine: Human consumption of ancient solar energy', *Climatic Change*, 61/1–2, 2003, pp. 31–44.

2 Bruno Latour, *Down to Earth: Politics in the New Climatic Regime*, trans. Catherine Porter, Cambridge: Polity, 2018, p. 82.

3 McKenzie Wark, *Molecular Red: Theory for the Anthropocene*, New York: Verso Books, 2015.

4 On the power of the counterfactual in the history of science, see the section 'Counterfactuals and the historian of science', *Isis*, 99, 2008, pp. 547–84. Its application to the history of technology remains to be explored.

5 Rolf Peter Sieferle, *The Subterranean Forest: Energy, Systems, and Industrial Revolution*, Cambridge: The White Horse Press, 2001.

6 Johan Philipp Butingen, *Sylva subterranea*, Gedruckt von C. Salfelden, 1693.

7 Gérard Borvon, *Histoire du carbone et du CO$_2$*, Paris: Vuibert, 2013, p. 119.

8 Jean Hassenfratz, *La Sidérotechnie ou l'art de traiter les minerais de fer pour en obtenir de la fonte, du fer, ou de l'acier*, Vol. 4, Paris: Firmin Didot, 1812.

9 E. A. Newell Arber, *The Natural History of Coal*, Cambridge: Cambridge University Press, 1911, p. 6.

10 Lissa Roberts and Joppe van Driel, 'The case of coal', in Lissa Roberts and Simon Werrett, eds, *Compound Histories: Materials, Governance and Production, 1760–1840*, Leiden: Brill, Cultural Dynamics of Science series, Vol. 2, 2017, pp. 57–84.

11 Henri Louis Duhamel du Monceau, *Éléments d'agriculture*, Vol. 1, Paris, 1762, pp. 182–6.

12 John Lindley and William Hutton, *Fossil Flora of Great Britain*, London: James Ridgway, 1831–3, pp. v–vi, cited by Roberts and van Driel, 'The case of coal', pp. 61–2.

13 Lewis Mumford, *Technics and Civilization*, London: Routledge & Kegan Paul, 1934, p. 156.

14 Andreas Malm, 'The origin of fossil capital: From water to steam in the British cotton industry', *Historical Materialism*, 21/1, 2013, pp. 15–68, p. 18.

15 Ibid.

16 David Landes, *The Unbound Prometheus: Technical Change and Industrial Development in Western Europe from 1750 to the Present*, Cambridge: Cambridge University Press, 2003 [1969].

17 Mumford, *Technics and Civilization*, p. 156.

18 Ibid., p. 157.

19 Malm, 'The origin of fossil capital'.

20 Edward Wrigley, *Energy and the English Industrial Revolution*, Cambridge: Cambridge University Press, 2010.

21 Mumford, *Technics and Civilization*, p. 157.

22 Ibid., p. 158.

23 Timothy Mitchell, *Carbon Democracy: Political Power in the Age of Oil*, London: Verso, 2011.

24 Malm, 'The origin of fossil capital', p. 17.

25 William Stanley Jevons, *The Coal Question: An Inquiry Concerning the Progress of the Nation and the Probable Exhaustion of the Coal Mines*, London: Macmillan and Co., 1865. This whole passage was inspired by Nuno Luis Madurerira, 'The anxiety of abundance: William Stanley Jevons and coal scarcity in the nineteenth century', *Environment and History*, 18, 2012, pp. 395–421.

26 A common distinction is made between resources (the known or estimated quantity of an energy or material) and reserves (the quantity that can be economically exploited with available technologies). See George Olah, *Beyond Oil and Gas: The Methanol Economy*, Weinheim: VHC, 2006, p. 27.

27 Jules Verne, *The Mysterious Island*, Chicago: Belford, Clarke & Co., 1884, p. 245.

28 Pierre Teissier, 'From the birth of fuel cells to the utopia of the hydrogen world', in Bernadette Bensaude-Vincent et al., eds, *Research Objects in Their Technological Setting*, Abingdon, Oxon: Routledge, 2017, pp. 70–86.

29 Marcellin Berthelot, 'Discours au banquet de la Chambre syndicale des produits chimiques, le 5 avril 1894', https://sniadecki.wordpress.com/1894/04/05/berthelot-01/.

30 Alain Gras, *Le Choix du feu: Aux origines de la crise climatique*, Paris: Fayard, 2007.

Chapter 11   *The bewitching power of oil*

1 Jevons, *The Coal Question*, pp. 2 and 197. Emphasis in original.

2 Mitchell, *Carbon Democracy*.

3 Matthieu Auzanneau, *Oil, Power, and War: A Dark History*, trans. John F. Reynolds and Richard Heinberg, London: Chelsea Green Publishing, 2018, p. 85.

4 Ibid., p. 167.

5 Rockefeller founded the University of Chicago in the 1890s, co-financed with Edmond de Rothschild the Henri-Poincaré Institute in 1928 in Paris and generously supported the emigration of Jews from territories conquered by Nazi Germany.

6 Mitchell, *Carbon Democracy*, Ch. 7; Auzanneau, *Oil, Power, and War*, Ch. 19.

7 Path dependence is a mechanism whereby a set of decisions at a given point in time is constrained by the systemic effects of decisions that have been made in the past.

8 David A. Kirsch, *The Electric Vehicle and the Burden of History*, New Brunswick: Rutgers University Press, 2000.

9 Edwin Black, *Internal Combustion: How Corporations and Governments Addicted the World to Oil and Derailed the Alternatives*, New York: St Martin's Griffin, 2007.

10 Olah, *Beyond Oil and Gas*, p. 6.

11 Auzanneau, *Oil, Power, and War*, p. 24.

12 Mitchell, *Carbon Democracy*, p. 36.

13 Céline Pessis, Sezin Topçu and Christophe Bonneuil, *Une autre histoire des 'Trente Glorieuses'*, Paris: La Découverte, 2013.

14 Isabelle Stengers and Philippe Pignarre, *Capitalist Sorcery: Breaking the Spell*, trans. and ed. Andrew Goffey, London: Macmillan, 2007.

15 Quanwei Chen et al., 'Investigating carbon footprint and carbon reduction potential using a cradle-to-cradle LCA approach on lithium-ion batteries for electric vehicles in China', *Journal of Cleaner Production*, 369, 2022, 133342.

16 Beginning of Ch. 10.

17 Hartmut Rosa, *Social Acceleration: A New Theory of Modernity*, trans. and introduced by Jonathan Trejo-Mathys, New York: Columbia University Press, 2013.

18 Rosa analyses the temporal dynamics of our societies by identifying three types of acceleration: technical acceleration constantly modifies our *habitus* and our environment; it is accompanied by an acceleration of the rhythm of life which

creates anxiety, exclusion and alienation; finally, it engenders the acceleration of social and cultural transformations which results in permanent instability and crises.

Chapter 12 The age of plastics

1 Richard Powers, *Gain*, London: Random House, 2001, p. 347.

2 Ibid., p. 352.

3 According to Pap N'Diaye, the number of researchers working in DuPont laboratories rose from 279 in 1927, to 687 in 1930, 847 in 1935 and 1,291 in 1940. *Du nylon et des bombes: DuPont, l'État et le marché, 1900–1970*, Paris: Belin, 2001.

4 A book making such accusations is H. C. Engelbrecht and F. C. Hanighen, *Merchants of Death: A Study of the International Armament Industry*, New York: Routledge, 1934.

5 This polyamide 6-6 did not secure the future of its inventor, as Carothers committed suicide in 1937. It is said that he was mainly concerned with confirming Staudinger's theory of macromolecules and considered himself a failure. He was often subject to depression.

6 Anecdote reported in Pap N'Diaye, 'La belle époque du nylon', *La Recherche*, 300, 1997, pp. 100–3.

7 An acid that can provide two positively charged hydrogen ions, i.e. two protons, in its reaction with a base.

8 An amine is a compound derived from ammonia (NH_3) in which at least one hydrogen atom bonded to the central nitrogen atom has been replaced by a carbon group (such as CH_3). A diamine is a compound containing two amine groups (such as $C_6H_{16}N_2$, hexamethylene diamine). It is said to be 'dry' when it is fully cross-linked and rigid.

9 See Susannah Handley, *Nylon: The Story of a Fashion Revolution*, Baltimore: Johns Hopkins University Press, 1999, pp. 18–19. The misfortunes of the first artificial silk may have inspired Alexander Mackendrick to make his film *The Man in the White Suit* (1950), which ridiculed synthetic textiles.

10 Cited from the *Detroit News* by Jeffrey Meikle, *American Plastic: A Cultural History*, New Brunswick, NJ: Rutgers University Press, 1995, p. 132.

11 DuPont presented *The Wonder World of Chemistry* at the Texas Centennial Exposition in 1936 and then exhibited nylon stockings at the San Francisco World's Fair in 1939. See David J. Rhees, 'Corporate advertising, public relations and popular exhibits: The case of DuPont', in Brigitte Schroeder-Gudehus, ed., *Industrial Society and Its Museums*, Paris: Cité des sciences et de l'industrie/Harwood Academic Publishers, 1993, pp. 67–76.

12 Handley, *Nylon*, pp. 46–51.

13 This culture was inaugurated with great success by Leo Baekeland, who, as early as 1907, manufactured a synthetic polymer called 'bakelite' which the firm promoted as 'the material with a thousand uses'. And it was indeed used to make telephones, radios, guns, billiard balls, coffee pots, jewellery, etc.

14 Plasticity differs from elasticity, which is to have no form or to return to its original form without change, while plasticity creates another identity.

15 Catherine Malabou, ed., *Plasticité*, Paris: Éditions Léo Scheer, 2000; Catherine Malabou, *Plasticity: The Promise of Explosion*, ed. Tyler Williams, Edinburgh: Edinburgh University Press, 2022.

16 William J. Hale, *Chemistry Triumphant*, Baltimore: The Williams & Wilkins Co., 1932.

17 Meikle, *American Plastic*.

18 Jeanne Guien, *Le Consumérisme à travers ses objets*, Paris: Éditions Divergences, 2021.

19 Handley, *Nylon*, p. 42.

20 Roland Barthes, *Mythologies*, trans. Annette Lavers, New York: Farrar, Straus & Giroux, 1972 [1957], p. 97.

21 As in Jaydee's 1993 song 'Plastic Dreams'.

22 Heather Davis, 'The domestication of plastic', *Interalia Magazine* (https://www.interaliamag.org/articles/heather-davis/). See also Heather Davis, *Plastic Matter*, Durham, NC: Duke University Press, 2022.

23 Barthes, *Mythologies*, p. 99.

24 Silica gel, for example, and especially polymerization agents, catalysts, cross-linking accelerators or inhibitors are introduced.

25 See Bernadette Bensaude-Vincent, *Éloge du mixte*, Paris: Hachette littératures, 1998, Ch. 10.

26 A. E. Standage and R. Prescott, 'High elastic modulus carbon fibre', *Nature*, 211, 1966, p. 169; Plastics & Rubber Institute, *Carbon Fibers: Technology, Uses and Prospects*, New York: Noyes Publications, 1986; J. Luyckx, 'Fibres de carbone', *Traité Matériaux non métalliques, Techniques de l'ingénieur*, Paris, 1994, pp. A2210-1–A2211-4.

27 See Ch. 6.

28 On the manufacture of carbon fibres, see Ch. 7. Just as synthetic polymers were initially a US speciality, carbon fibres were a Japanese speciality. Carbon fibre exports have helped to spread the Japanese culture, if not the Japanese business model, around the world.

29 François Dagognet, *Rematérialiser: Matières et matérialismes*, Paris: Vrin, 1985, p. 125.

30 Ibid., pp. 119 and 126.

31 Janine Benyus, *Biomimicry: Innovation Inspired by Nature*, New York: William Morrow, 1997.

32 For example, in car bumpers or in the rotor blade system of helicopters, the substitution of composites for metal has made it possible to integrate several functions into a single part.

33 For this section, we are very much indebted to the work of Baptiste Monsaingeon (*Homo detritus: Critique de la société du déchet*, Paris: Seuil, 2017, especially Ch. 3, 'Un monde plastique').

34 For the Atlantic, see E. J. Carpenter and K. L. Smith, 'Plastics on the Sargasso Sea surface', *Science*, 175, 1972, pp. 1240–1. For the Pacific, see E. L. Venrick et al., 'Man-made objects on the surface of the central North Pacific Ocean', *Nature*, 241, 1973, p. 271.

35 Charles Moore, 'Across the Pacific Ocean, plastics, plastics, everywhere', *Natural History Magazine*, 112/9, 3 November 2003. Cited by Baptiste Monsaingeon, '"Oceans of plastic": Heterogeneous narrations of an ongoing disaster', *LIMN*, 3 (https://limn.it/articles/oceans-of-plastic-heterogeneous-narrations-of-an-ongoing-disaster/).

36 Jenna R. Jambeck et al., 'Plastic waste inputs from land to the ocean', *Science*, 347, 13 February 2015, pp. 768–71.

37 Michel Serres, *Genesis*, trans. Geneviève James and James Nelson, Ann Arbor: University of Michigan Press, 1995 [1982], p. 1.

38 https://www.septiemecontinent.com/.

39 Baptiste Monsaingeon, 'Sur les traces des océans de plastique: Faire monde avec l'irréparable', *Techniques et Culture*, 65–6, 2016, pp. 34–47; see also Baptiste Monsaingeon, 'Plastiques: Ce continent qui cache nos déchets', *Mouvements*, 87, 2016, pp. 48–58.

40 Levi, *The Periodic Table*, pp. 143–4.

Chapter 13 Working towards a more sustainable economy

1 In biology, anabolism refers to all the synthesis reactions that take place in an organism (protein synthesis, tissue growth and renewal, etc.), while catabolism refers to all the degradation reactions (energy consumption, digestion, cell death, etc.). Anabolism and catabolism are intimately linked by the organism and constitute the two essential and complementary components of its metabolism.

2 This second option will be the subject of Ch. 14.

3 Marc R. Finlay, 'Old efforts at new uses: A brief history of chemurgy and the American search for biobased materials', *Journal of Industrial Ecology*, 7/3–4, 2003, pp. 33–46.

4 Nicholas Georgescu-Roegen, *The Entropy Law and the Economic Process*,

Cambridge, MA: Harvard University Press, 1971. René Passet, another bioeconomy advocate, defines it as 'an approach that opens the economy to the biosphere, of which it is only a subsystem'. René Passet, *Les Grandes Représentations du monde et de l'économie à travers l'histoire*, Paris: Les Liens qui libèrent, 2010, p. 896.

5 Nicholas Georgescu-Roegen, 'Energy and economic myths', *Southern Economic Journal*, 41/3, 1975, pp. 347–81, p. 371.

6 Roland Verhé, 'Dilemma: Petrochemistry and oleochemistry as resources for fuel and oleochemicals', *European Journal of Lipid Science and Technology*, 112/4, 2010, pp. 427–47, p. 427.

7 Martino Nieddu et al., 'Existe-t-il vraiment un nouveau paradigme de la chimie verte?', *Natures Sciences Sociétés*, 22, 2014, pp. 103–13.

8 See Ch. 5 above.

9 On the tension between these two concepts of bio-economy, see Les Levidow, Kean Birch and Theo Papaioannou, 'EU agri-innovation policy: Two contending visions of the bio-economy', *Critical Policy Studies*, 6/1, 2012, pp. 40–65.

10 https://www.oecd.org/futures/long-termtechnologicalsocietalchallenges/thebio economyto2030designingapolicyagenda.htm.

11 Ibid.

12 Kaushik Sunder Rajan, *Biocapitalism*, Durham, NC: Duke University Press, 2006.

13 Pierre Monsan, 'De la science vers la recherche industrielle dans le domaine des biotechnologies blanches, en France et à l'étranger', in 'Biotechnologies blanches et biologie de synthèse', report of the Académie des technologies, tabled 9 July 2014 (https://www.academie-technologies.fr/publications /biotechnologies-blanches-et-biologie-de-synthese-rapport-2015/).

14 CRISPR stands for Clustered Regularly Interspaced Short Palindromic Repeats, Cas9 for the Crispr-Associated 9 protein.

15 George Church and Ed Regis, *Regenesis: How Synthetic Biology Will Reinvent Nature and Ourselves*, New York: Basic Books, 2012.

16 Ibid., p. 4.

17 Steven A. Benner, Zuyni Yang and Fei Chen, 'Synthetic biology, tinkering biology and artificial biology. What are we learning?', *Comptes Rendus Chimie*, 14/4, 2011, pp. 372–87.

Chapter 14 *The carbon market*

1 '*Haec carbone notasti*' we read in the *Satires* of Persius (V. 108): 'you have blamed this', literally, you have marked or notified it with charcoal. To approve or acquit, they said '*creta notare*', 'to note with chalk'.

2 https://icapcarbonaction.com/en/publications/emissions-trading-worldwide
-icap-status-report-2015, p. 3.

3 Through the United Nations Framework Convention on Climate Change
(UNFCCC), which verifies the validity of carbon projects and issues emission
reduction credits.

4 This idea goes back to the work of John Dales, an environmental economist
who invented the concept of 'pollution rights markets', and Ronald Coase,
winner of the Nobel Prize in Economics in 1991. According to the latter, 'a
polluter should not be seen as doing something bad and should be prevented
from doing so [. . .]. Pollution does bad things as well as good things. People do
not pollute because they like to pollute. They do it because it is a cheaper way
to produce something else. Producing cheaper is good; losing value because of
pollution is bad. You have to compare the two, that's the right way to look at
it.' Thomas W. Hazlett, 'Looking for results: An interview with Ronald Coase
(Nobel laureate Ronald Coase on rights, resources, and regulation)', *Reason
Magazine – Free minds and free market*, January 1997 (https://reason.com/1997
/01/01/looking-for-results/).

5 As transportation systems are exempted from the institutional carbon market,
the SNCF [the French railways], for example, whose 'eco-calculator' always
displays the CO_2 footprint of each ticket, does not offset its emissions itself
but offers it to its users on a voluntary basis, via a partnership with the offset
provider ActionCarbone. Later on, the train company set an internal carbon
price for its suppliers.

6 BlueNext, co-founded in 2007 by the transatlantic group of financial market
companies New York Stock Exchange-Euronext and the Caisse des Dépôts,
closed in 2012 following a massive VAT fraud scandal and the European
Commission's refusal to entrust it with auctioning CO_2 allowances.

7 A rate that compares a future income or expense with an immediate income
or expense. See David Weisbach and Cass R. Sunstein, 'Climate change and
discounting the future: A guide for the perplexed', *Yale Law & Policy Review*,
27, 2008, pp. 433–57.

8 Isabelle Delpla, *Du Pays vide: Réfuter le solipsisme politique*, Paris: Vrin, 2023.

9 Donald MacKenzie, 'Making things the same: Gases, emission rights and the
politics of carbon markets', *Accounting, Organizations and Society*, 34/3, 2009,
pp. 440–55.

10 See Ch. 2.

11 See Ch. 17.

12 See Ch. 5, where carbon also begins its chemical career as 'one among many'.

13 See Ch. 16.

14 Laurence Raineau, *L'Utopie de la monnaie immatérielle*, Paris: PUF, 2004. See also Laurence Duchêne and Pierre Zaoui, *L'Abstraction matérielle: L'argent au-delà de la morale et de l'économie*, Paris: La Découverte, 2012.

15 Dan Welch, co-editor of *Ethical Consumer Magazine*, quoted by Austen Naughten, 'Designed to fail? The concepts, practices and controversies behind carbon trading', *FERN*, 2010, p. 12 (https://www.fern.org/fileadmin/uploads /fern/Documents/FERN_designedtofail_internet_0.pdf).

16 Augustin Fragnière, *La Compensation carbone: Illusion ou solution?*, Paris: PUF, 2009.

17 David Adam, 'Can planting trees really give you a clear carbon conscience?', *The Guardian*, 7 October 2006; Kevin Smith, *The Carbon Neutral Myth: Offsets Indulgences for Your Climate Sins*, Amsterdam: Carbon Trade Watch, 2007.

18 Larry Lohmann, 'Carbon trading, climate justice and the production of ignorance: Ten examples', *Development*, 51/3, 2008, pp. 359–65.

19 Académie des technologies, 'Quel prix de référence du CO_2?', opinion and report from the Académie des technologies, January 2017.

20 Currently, eighteen states have implemented some form of carbon taxation in varying ways. Rates range from €118/ tCO2-eq (Sweden) to less than €0.9/ tCO2-eq (Mexico). Revenue redistribution options range from contributions to the general budget (Sweden) to grants for energy efficiency measures, lower employer contributions (Denmark) and health insurance funding (Switzerland). There are numerous exemptions, both partial and total, particularly for companies exposed to strong competition (agriculture, fisheries), or those already subject to the cap-and-trade system of the institutional carbon market. In France, the Yellow Vest uprising of 2018 is often said to have been caused by the unfairness of a carbon tax that penalized the middle class and peri-urban populations who have to make long car journeys to work. However, it should be pointed out that it was not a 'real' (i.e. universal) carbon taxation that was at issue, but simply a fuel tax at the pump, resulting from the increase in a domestic consumption tax on energy products. On the contrary, a real carbon tax would have enabled a fairer distribution of revenues, without putting most of the burden on domestic end consumers only.

21 Christian de Perthuis and Pierre-André Jouvet, 'L'accord sur le climat devra étendre le marché carbone à l'échelle mondiale', *Le Monde*, 16 September 2015. See the response in the form of a warning from Jessica F. Green, 'Don't link carbon markets', *Nature*, 543, 2017, pp. 484–6.

22 Justin Leroux, Étienne Billette de Villemeur and Myriam Jézéquel, 'La "dette carbone", une idée novatrice pour sortir de l'impasse climatique', *Gestion*, 40/4,

2015 (https://www.revuegestion.ca/la-dette-carbone-une-idee-novatrice-pour-sortir-de-l-impasse-climatique).

23 Fragnière, *La Compensation carbone*, Introduction. See also Larry Lohmann, ed., *Carbon Trading: A Critical Conversation on Climate Change, Privatization and Power*, Uppsala: Dag Hammarskjöld Foundation, 2006.

24 On the critique of technical fixes, see Michael Huesemann and Joyce Huesemann, *TechNO-Fix: Why Technology Won't Save Us or the Environment*, Gabriola Island: New Society Publishers, 2011; Evgeny Morozov, *To Solve Everything, Click Here: The Aberration of Technological Solutionism*, New York: PublicAffairs, 2013; Naomi Klein, *This Changes Everything: Capitalism vs the Climate*, New York: Simon & Schuster, 2014.

Chapter 15 Carbon cosmogony

1 The 'Big Bang' is not an explosion. It is an ironic term introduced in 1950 by the British astrophysicist and cosmologist Fred Hoyle to mock the set of theories he opposed at the time, which described the currently observable universe as having expanded rapidly from a dense, hot state.

2 In the form of helium-3 (2 protons + 1 neutron) and helium-4 (2 neutrons + 2 protons).

3 Deuterium is the isotope-2 of hydrogen: 1 proton + 1 neutron.

4 The hydrogen atom has a spontaneous electron spin reversal probability of once every eleven million years.

5 Here we are meeting an 'altruistic' profile of carbon – both conductor and facilitator, both central and able to step aside to make room for other elements – which could be relevant to the discussion of 'selfish carbon' to be broached in Ch. 16.

6 Hans A. Bethe, 'Energy production in stars', *Physical Review*, 55, 1939, pp. 434–56.

7 Ibid., p. 446.

8 Carbon is the fourth element (0.5% by mass) after hydrogen (73.9%), helium (24%) and oxygen (1%). After carbon come neon, iron and nitrogen (0.1% each), then silicon (0.07%).

9 Ernst Öpik, 'Stellar models with variable composition. II: Sequences of models with energy generation proportional to the fifteenth power of temperature', *Proceedings of the Royal Irish Academy A*, 54, 1951, pp. 49–77. The paper by this Estonian astronomer, who had fled from the Red Army to Germany and then Northern Ireland, went unnoticed by British astrophysicists at the time. Edwin E. Salpeter, 'Nuclear reactions in stars without hydrogen', *Astrophysical Journal*, 115, 1952, pp. 326–8.

10 Simon Mitton, *Conflict in the Cosmos: Fred Hoyle's Life in Science*, Washington, DC: Joseph Henry Press, 2005.

11 Charles W. Cook et al., 'B^{12}, C^{12}, and the red giants', *Physical Review*, 107, 1957, pp. 508–15.

12 Fred Hoyle, *Galaxies, Nuclei, and Quasars*, New York: Harper & Row, 1965, p. 147.

13 Brandon Carter, 'Large number coincidences and the anthropic principle in cosmology', in M. S. Longair, ed., *Confrontation of Cosmological Theories with Observational Data: Proceedings of the Symposium, Krakow, Poland, September 10–12*, Dordrecht: Reidel Publishing Co., 1974, pp. 291–8.

14 Helge Kragh, 'An anthropic myth: Fred Hoyle's carbon-12 resonance level', *Archive for the History of Exact Sciences*, 64/6, 2010, pp. 721–51.

15 The only way would be to assume an evolution of the laws of nature, i.e. a variability of invariants, but at what cost to physics? 'Such a conception has no chance of ever being adopted by scientists; in the sense in which they would understand it, they could not adhere to it without denying the legitimacy and the very possibility of science,' wrote Henri Poincaré in 1911 (in 'L'évolution des lois', *Scientia*, IX, 1911, pp. 275–92, p. 275). Moreover, if such variability were to be proven, the anthropic significance of Hoyle's prediction would lose considerable force.

16 Fred Hoyle, 'Some remarks on cosmology and biology', *Memorie della Societa Astronomica Italiana*, 62, 1991, pp. 513–18, p. 518.

17 See Ch. 4.

18 See Ch. 6.

19 On carbon in the planet formation and early differentiation, see Beth N. Orcutt, Isabelle Daniel and Radjeep Dasgupta, *Deep Carbon: Past to Present*, Cambridge: Cambridge University Press, 2020.

20 Able to develop without taking organic molecules from the environment by drawing their energy from light (photoautotrophs) or from the reduction of inorganic substrates (chemoautotrophs).

21 Jan Zalasiewicz and Mark Williams, *The Goldilocks Planet: The Four Billion Year Story of Earth's Climate*, Oxford: Oxford University Press, 2012.

22 David Archer, *The Global Carbon Cycle*, Princeton: Princeton University Press, 2010. See also: https://www.coursera.org/lecture/global-warming/the-weathering-co2-thermostat-rg7tn.

23 Phytoplankton produce half of the oxygen consumed by living things and volcanoes. Volcanoes consume oxygen by burning organic carbon compounds in sediments in subduction zones.

24 Pablo Jensen, *Des atomes dans mon café crème: La physique peut-elle tout expliquer?*,

Paris: Seuil, 2004, Ch. XXIII, 'Le tout et les parties, et réciproquement'. See also Thierry Martin, ed., *Le Tout et les parties dans les systèmes naturels*, Paris: Vuibert, 2007.

Chapter 16 Turbulence in the biosphere

1 See the previous chapter.

2 Richard Dawkins, *The Selfish Gene*, Oxford: Oxford University Press, 1976.

3 Jean-Baptiste Dumas, *Essai de statique chimique des êtres organisés*, Paris: Fortin-Masson, 1842. Quoted in the new edition, Brussels: Éditions Culture et civilisation, 1972, p. 192. On Dumas's conception of the relation between life and chemistry, see Ch. 5.

4 For Roston, see Prologue, note 8; for Levi and Zalasiewicz, see Prologue, note 13.

5 See Ch. 6.

6 James Lovelock, 'Atmospheric homeostasis by and for the biosphere: The Gaia hypothesis', *Tellus*, 26/1–2, 1974, pp. 2–9.

7 Tyler Volk, *CO$_2$ Rising: The World's Greatest Environmental Challenge*, Cambridge, MA: MIT Press, 2008.

8 Mikhail Budyko, quoted in Noah B. Bornheim, 'History of climate engineering', *Wiley Interdisciplinary Reviews*, 1/6, 2010, pp. 891–7, p. 894.

9 President's Science Advisory Committee, 'Restoring the quality of our environment'. See James Rodger Fleming, *Fixing the Sky: The Checkered History of Weather and Climate Control*, New York: Columbia University Press, 2010, p. 238.

10 National Academy of Sciences et al., *Policy Implications of Greenhouse Warming: Mitigation, Adaptation, and the Science Base*, Washington, DC: National Academy Press, 1992, Ch. 28, 'Geoengineering', pp. 433–64.

11 The name was coined in memory of Thomas H. Huxley and Cesare Emiliani, who identified these coccoliths in layers of marine sediment in the nineteenth century.

12 A bloom of coccolithophores in Douarnenez Bay analysed by the REPHY and REPHYTOX missions (Réseau d'observation et de surveillance du phytoplancton et des phycotoxines) of the IFREMER (Institut français de recherche pour l'exploitation de la mer) on 30 September 2013 revealed a concentration of over six million coccolithophores per litre.

13 It is estimated that 1.5 million tonnes of calcite are deposited on the seabed every year.

14 Coccolith calcite is both solid and brittle, as it has not undergone the intense compression caused by the accumulation of soil over geological time. This is

why it has long been prized as a building material (ashlar) and writing material (white chalk for blackboards). Today's chalk is produced mainly from plaster, a derivative of gypsum, an entirely crystalline calcium sulphate.

15 FAO (Food and Agriculture Organization of the United Nations), 'Agriculture's greenhouse gas emissions on the rise', April 2014 (https://www.fao.org/news /story/en/item/216137/icode/).

16 Here is an estimate of the carbon stocks (in tonnes of CO_2) of a forest per hectare based on a concrete case: above-ground biomass: 582 tCO_2, woody biomass: 727 tCO_2, carbon from underground biomass: 145 tCO_2 (http://www .observatoire-comifac.net/docs/confCarbon/2010brazzaville/Jour2/Bayol_FRM _CarboneConcession.pdf).

17 'Les sols, des milieux vivants très fragiles', Suds en Ligne: Les dossiers thématiques de l'IRD (https://www.doc-developpement-durable.org/file/eau/lutte -contre-erosion_protection-sols/SolsMilieuxTresFragiles.pdf).

18 Charles Darwin, *The Formation of Vegetable Mould Through the Action of Worms*, London: John Murray, 1881.

19 ANR Collective, 'Réflexion systémique sur les enjeux et méthodes de la géo-ingénierie de l'environnement', Report of the REAGIR Foresight Workshop, May 2014.

20 Justus von Liebig, *Chimie appliquée à la physiologie végétale et à l'agriculture*, Paris: Fortin et Masson, 1844. See also Marika Blondel Megrelis, *La Chimie agricole de Liebig*, Paris: Lavoisier, 2017.

21 Another solution being considered involves bioengineering: manufacturing chemotrophic bacteria (which reduce CO_2 directly without photosynthesis) and integrating them deep underground. Bacteria – as well as archaea – have shown an extraordinary ability to survive in extreme conditions of temperature (in hydrothermal springs) or pressure (at a depth of 5,000 m).

22 See Ch. 10.

23 Christophe Naisse, 'Potentiel de séquestration de carbone des biochars et hydrochars, et impact après plusieurs siècles sur le fonctionnement du sol', Sciences de la Terre, thèse de l'université Pierre-et-Marie-Curie, Paris 6, 2014; HAL Id: tel-01130038.

24 Y. Kuzgakov, J. K. Friedel and K. Stahr, 'Review of mechanisms and quantification of the priming effect', *Soil Biology and Biochemistry*, 32, 2000, pp. 1485–98.

25 E. Blagodatskaya and Y. Kuzgakov, 'Mechanisms of real and apparent priming effects and their dependence on soil microbial biomass and community structure: Critical review', *Biology and Fertility of Soils*, 45, 2008, pp. 115–31.

26 Bertrand Guenet et al., 'Priming effect: Bridging the gap between terrestrial and aquatic ecology', *Ecology*, 91/10, 2010, pp. 2850–61.

Chapter 17 Rethinking time with carbon

1 Paul J. Crutzen and Eugene F. Stoermer, 'The "Anthropocene"', *Global Change Newsletter*, 41, 2000, pp. 17–18.

2 In this case, the Anthropocene would refer less to an epoch than to a catastrophe in the mathematical sense (a sudden transition that causes a system to split), or to what geologists call a boundary, such as the K–Pg boundary (Kreide–Paläogen). This was the moment of the mass extinction of the dinosaurs, marking the end of the Cretaceous and the beginning of the Palaeogene, and can be read in the soil by a characteristic geological signature (a thin layer of clay with a high iridium content on top of a chalky layer). As we saw in Ch. 15, the earth's history is palaeo-biological; it is linked to the great extinctions.

3 An international working group on the Anthropocene set up within the International Commission on Stratigraphy has been working on this issue for over ten years. At the 35th International Geological Congress, in 2016, it endorsed the recommendation to include the Anthropocene as a geological epoch within geological chronology.

4 See Ch. 1.

5 The calcination of limestone converts $CaCO_3$ into lime, CaO, releasing CO_2 into the atmosphere. Although this emission is partially offset by the absorption of CO_2 during the setting of the cement, cement manufacture accounts for 7 to 8% of man-made CO_2 emissions, as the calcination process consumes a great deal of fossil energy.

6 The closest natural analogue to man-made emissions is degassing from the earth, which is estimated at 0.1 GtC/year, a hundred times less than emissions caused by human activities.

7 Aant Elzinga, 'Polar ice cores: Climate change messengers', in Bensaude-Vincent et al., eds, *Research Objects in Their Technological Setting*, pp. 215–31.

8 Paul J. Crutzen, 'The geology of mankind', *Nature*, 415, 3 January 2002, p. 23.

9 In 2013, publications on geoengineering in humanities and social science journals even outnumbered those in natural science journals (Björn-Ola Linnér and Victoria Wibeck, 'Dual high-stake emerging technologies: A review of the climate engineering research literature', *WIREs Climate Change*, 6, 2015, pp. 255–68).

10 Alvin Toffler, *Future Shock*, New York: Random House, 1970. Fifteen million copies have been sold in English alone.

11 Rosa, *Social Acceleration*; Hartmut Rosa, *Alienation and Acceleration: Towards a Critical Theory of Late-modern Temporality*, Natchitoches, LA: NSU Press, 2010; see also Ch. 11.

12 Judy Wacjman, *Pressed for Time: The Acceleration of Life in Digital Capitalism*, Chicago: University of Chicago Press, 2014.

13 Jan Zalasiewicz et al., 'The Anthropocene: A new epoch of geological time?', *Philosophical Transactions of the Royal Society of London A*, 369, 2011, pp. 835–41.

14 Over the past decade, the work conducted by the Anthropocene Working Group (AWG), a subcommunity of geologists in charge of assessing the stratigraphic evidence of the Anthropocene, has been closely accompanied by the humanities (Christoph Rosols, Georg Schäfer and Bernd Scherer, 'Evidence and experiment: Curating contexts of Anthropocene geology', *Anthropocene Review*, 10, 2023, pp. 330–9).

15 Simon L. Lewis and Mark A. Maslin, 'Defining the Anthropocene', *Nature*, 519, 12 March 2015, pp. 171–80.

16 See Ch. 12.

17 Maurice Fontaine, *Rencontres insolites d'un biologiste autour du monde*, Paris: L'Harmattan, 1999, pp. 36–7.

18 See Chs 10 and 11.

19 Jason Moore, *Anthropocene or Capitalocene? Nature, History, and the Crisis of Capitalism*, New York: Verso Books, 2015.

20 Christophe Bonneuil and Jean-Baptiste Fressoz, *L'Événement anthropocène: La Terre, l'histoire et nous*, Paris: Seuil, 2014.

21 See Victor Petit, 'Technocène?', séminaire Digital Studies, Institut de recherche et d'innovation du Centre G. Pompidou, Paris, 10 February 2016, Session 5. 'Toward a general ecology' (https://digital-studies.org/wp/digital-studies -seminar-2014-2015/); Victor Petit, 'La mésologie à l'épreuve du technocène', in Marie Augendre, Jean-Pierre Llored and Yann Nussaume, eds, *La mésologie: Un autre paradigme pour l'anthropocène? Autour et en présence d'Augustin Berque*, Paris: Hermann, 2018, pp. 101–7.

22 Sébastien Dutreuil, 'Gaïa, hypothèse de recherche pour le système Terre ou philosophie de la nature?', thèse de l'université Paris 1 Panthéon-Sorbonne, 2016, pp. 604–8.

23 Jean-Baptiste Fressoz, 'L'Anthropocène et l'esthétique du sublime', *Mouvements*, 16 September 2016 (https://mouvements.info/sublime-anthropocene/).

24 Jan Zalasiewicz et al., 'The new world of the Anthropocene', *Environmental Science & Technology*, 44, 2010, pp. 2228–31, p. 2231.

25 Crutzen and Stoermer, 'The "Anthropocene"', p. 17.

26 Pierre de Jouvancourt and Christophe Bonneuil, 'En finir avec l'épopée: Grand récit historique de l'environnement global, géopouvoir et fabrication de subjectivités dans l'Anthropocène', in Émilie Hache, ed., *De l'univers clos au monde infini*, Paris: Éditions Dehors, 2015, pp. 57–105.

27 Catherine Larrère and Raphaël Larrère, 'Les transitions écologiques à Cerisy', *Natures, Sciences & Sociétés*, 24, 2016, pp. 242–50, p. 249. See also their 'Peut-on échapper au catastrophisme?', in Dominique Bourg, Pierre-Benoît Joly and Alain Kaufmann, eds, *Du risque à la menace: Penser la catastrophe*, Paris: PUF, 2013, pp. 199–216.

28 Michael S. Northcott, *A Political Theology of Climate Change*, Grand Rapids, MI: Eerdmans Publishing, 2014; Lydia Barnett, 'The theology of climate change: Sin as agency in the Enlightenment's Anthropocene', *Environmental History*, 20/1, 2015, pp. 217–37.

29 The focus on carbon sequestration can be seen from a comparison between the roadmap for the Rio 2 Summit in 2012 (F. Biermann et al., 'Science and government: Negotiating the Anthropocene: Improving Earth system governance', *Science*, 335/6074, March 2012, pp. 1306–7) and that for the Paris Summit in 2015 (Kevin Anderson, 'Talks in the city of light generate more heat', *Nature*, 528, 31 December 2015, p. 437).

30 Gerard 't Hooft and Stefan Vandoren, *Time in Powers of Ten*, Singapore: World Scientific Publishing, 2014.

31 Richard Monastersky, 'Anthropocene: The human age', *Nature*, 519, 11 March 2015, pp. 144–7, p. 415.

32 Jacques Le Goff, 'Merchant's time and church time in the Middle Ages' (pp. 29–42) and 'Labor time in the "crisis" of the fourteenth century: From medieval time to modern time' (pp. 43–52), in *Time, Work, and Culture in the Middle Ages*, trans. Arthur Goldhammer, Chicago: University of Chicago Press, 1980; Helga Nowotny, *Time: The Modern and Postmodern Experience*, trans. Neville Plaice, Cambridge: Polity, 1994; Peter Galison, *Einstein's Clocks and Poincaré's Maps*, New York: Norton & Company, 2004; Kevin Birth, *Objects of Time: How Things Shape Temporality*, London: Palgrave Macmillan, 2012.

33 Norbert Elias, *Time: An Essay*, trans. Edmund Jephcott, Oxford: Blackwell, 1992, pp. 8–9.

34 Henri Bergson, *Creative Evolution*, trans. Arthur Mitchell, New York: Henry Holt and Company, 1911, p. xi.

35 Lucretius, *De Rerum Natura*, ed. William Ellery Leonard (http://www.perseus .tufts.edu/hopper/text?doc=Perseus:abo:phi,0550,001:2:2).

36 Taking up Peter Sloterdijk's critique of 'spheres', Bruno Latour, in *Facing Gaia: Eight Lectures on the New Climatic Regime* (trans. Catherine Porter, Cambridge: Polity, 2017), points out that 'notions of globe and global thinking include the immense danger of unifying too quickly what first needs to be *composed*' (p. 138, emphasis in original). He suggests replacing the image of the globe with the work of loops: 'After each passage through a loop, we become *more sensitive*

and *more reactive* to the fragile envelopes that we inhabit' (p. 140, emphasis in original).

37 Bruno Latour, *We Have Never Been Modern*, trans. Catherine Porter, Cambridge, MA: Harvard University Press, 1993; Latour, *Facing Gaia*.

38 Cthulhu is an extraterrestrial chimera with dragon wings and cuttlefish tentacles coined by science fiction writer H. P. Lovecraft ('The call of Cthulhu', *Weird Tales*, February 1928). Haraway gives it a twist: she modifies its spelling by moving the 'h' (Chthulu) to name a symbiosis of chthonic figures, from Gaia to Mami Wata, a 'sym-chthonic' figure.

39 Donna Haraway, 'Anthropocene, Capitalocene, Plantationocene, Chthulucene: Making kin', *Environmental Humanities*, 6, 2015, pp. 159–65, p. 160.

40 Among historians, the idea of a single chronology of events has not become the only solution for thinking about a 'universal history' (see Helge Jordheim, 'Synchronizing the world: Synchronism as historiographical practice, then and now', *History of the Present*, 7, 2017, pp. 59–95).

41 See Ch. 14.

42 Esther Turnhout and Susan Boonman-Berson, 'Databases, scaling practices and the globalization of biodiversity', *Ecology and Society*, 16/1, 2011, art. 35 (http://www.ecologyandsociety.org/vol16/iss1/art35/); Vincent Devictor and Bernadette Bensaude-Vincent, 'From ecological records to big data: The invention of global biodiversity', *History and Philosophy of the Life Sciences*, 38/13, 2016, pp. 11–23. Bernadette Bensaude-Vincent, 'Rethinking time in response to the Anthropocene: From timescales to timescapes', *The Anthropocene Review*, 9/2, 2021, pp. 206–19.

43 Anna Lowenhaupt Tsing, 'On nonscalability: The living world is not amenable to precision-nested scales', *Common Knowledge*, 18/3, 2012, pp. 505–24.

44 Bernadette Bensaude-Vincent, 'Slow–fast: Un faux débat?', *Natures, Sciences et Société*, 22/3, 2014, pp. 254–61.

45 William McDonough, *Cradle to Cradle: Remaking the Way We Make Things*, New York: North Point Press, 2002.

46 Archer, *The Global Carbon Cycle*, Ch. 1.

47 Haraway, 'Anthropocene, Capitalocene', p. 161.

48 See Ch. 16.

49 Archer, *The Global Carbon Cycle*.

50 See Ch. 15.

51 Archer, *The Global Carbon Cycle*, pp. 15–16.

52 Ibid., pp. 177–8.

53 Anna Lowenhaupt Tsing, *The Mushroom at the End of the World*, Princeton: Princeton University Press, 2017.

54 Alexandra Arènes, Bruno Latour and Jérôme Gaillardet, 'Giving depth to the surface: An exercise in the Gaia-graphy of critical zones', *The Anthropocene Review*, 5/2, 2018, pp. 120–35; Bruno Latour and Pieter Wiebel, *Critical Zones: The Science and Politics of Landing on Earth*, Cambridge, MA: MIT Press, 2020.

55 Gilbert Simondon, *On the Mode of Existence of Technical Objects*, trans. Cécile Malaspina and John Rogove, Minneapolis: University of Minnesota Press (Univocal), 2016 [1958].

Epilogue: The heteronyms of carbon

1 Richard Smalley, 'Discovering the fullerenes', Nobel lecture, 1996, p. 90 (https://www.nobelprize.org/prizes/chemistry/1996/smalley/lecture/). Our emphasis.

2 Dagognet, *Rematérialiser*; Jessica Riskin, *The Restless Clock: A History of Centuries-Long Argument Over What Makes Living Things Tick*, Chicago: University of Chicago Press, 2015; Sarah Ellenzweig and John H. Zammito, eds, *The New Politics of Materialism: History, Philosophy, Science*, London: Routledge, 2017.

3 Jane Bennett, *Vibrant Matter: A Political Ecology of Things*, Durham, NC: Duke University Press, 2010.

4 Peter Frazl et al., eds, *Active Materials*, Berlin/Boston: De Gruyter, 2021.

5 See Ch. 4.

6 While the first part of Simondon's *On the Mode of Existence of Technical Objects* focuses on technical objects, the third part unfolds, differentiates and converges a multiplicity of modes of being-in-the-world: magic, religion, technique, aesthetics, theory, practice, rite, dogma, ethics, science, organization, politics and philosophy.

7 See Prologue, note 15.

8 Latour, 'Reflections on Étienne Souriau's *Les différents modes d'existence*', p. 312.

9 See Prologue, note 15.

10 Michael Lynch, 'Ontography: Investigating the production of things, deflating ontology', *Social Studies of Science*, 43/3, 2013, pp. 444–62.

11 For example, the ontography practised in archaeology is a thought experiment that enables archaeologists to produce descriptions of things in ontological coordinates other than their own, such as animistic ones (Martin Holbraad, 'Ontology, ethnography, archaeology: An afterword on the ontography of things', *Cambridge Archaeological Journal*, 19, 2009, pp. 431–41).

12 Graham Harman, *The Quadruple Object*, New York: Zero Books, 2011. This object-oriented ontology (OOO) includes an 'ontography' consisting of a graphic representation (inspired by a standard pack of cards) of all the possible

relationships (tension, fusion and fission) between the four poles of reality according to Harman (real objects, real qualities, sensible objects, sensible qualities).

13 Ian Bogost, *Alien Phenomenology, or What It's Like to Be a Thing*, Minneapolis: University of Minnesota Press, 2012, p. 38. Emphasis in original.

14 Humberto R. Maturana, *Biology of Cognition*, Urbana: University of Illinois Press, 1970.

15 In *Of Grammatology* (trans. Gayatri Chakravorty Spivak, Baltimore: Johns Hopkins University Press, 1976), Jacques Derrida called this precisely 'arche-writing', drawing on the work of the palaeoanthropologist André Leroi-Gourhan, for whom cave paintings constituted a form of writing that predated linear, phonetic writing. In so doing, Derrida intended to open up the possibility of a reversal that would signal the deconstruction of the Western *logos*. Our ontography in no way aims to deconstruct *logos* or exclude *anthropos*. On the contrary, it accepts the fact that we are the storytellers, and welcomes as many humans and *logoï* as we need to tell good stories. The more the merrier . . .

16 This is due to a most instructive etymological confusion, linked to the term *scruple*. A *scrupulus* is a small *scrupus*, a pointed stone that would get lodged in the sandals of Roman legionaries, hindering their marching, hence the modern meaning of scruple as something that troubles the mind by preventing it from moving forward. But *scrupulus* also referred to a small stone weighing twenty-four grains or one twenty-fourth of an ounce, and therefore a unit of measurement, approximately equal to one gram today.

17 Levi, *The Periodic Table*, p. 227.

18 Ibid., p. 234.

19 https://all-geo.org/metageologist/2013/12/story-of-an-atom-birth-to-earth/.

20 https://prezi.com/ykbnbcynxkk6/copy-of-history-of-a-carbon-atom/.

21 Evelyn Mervine, geologist: 'My life started out easy, in the air. I had the pleasure of being in a molecule with two extremely nice oxygen atoms' (https://blogs.agu.org/georneys/2011/05/28/blast-from-the-past-carbon-cycle-story/); Trevone Vassell, in an online illustrated children's book: 'This is the story of my cycle. I never disappear or get created. I just change forms' (https://www.storyjumper.com/book/read/6293182/The-Life-Of-A-Carbon-Atom), etc.

22 *Pessoa* means 'person' in Portuguese.

Index

abduction, and the anthropic principle 187–8

acetylene 53

acids 53, 54

affordances in carbon chemistry 70–1, 72, 74–5, 96, 222, 223
coal 123–5

aggregated diamond nanorods 109

Agnano, Lake, *Grotta del Cane* 17, 18

agriculture
and the bioeconomy 154–5
improving the soil 201–2, 203

agroforestry 203

aircraft manufacture 147, 148

alchemy 34, 72

alcohols 53–4

aldehydes 53, 54

Alder, Kurt 2

alizarin 72

alkalis 54

alkaloids 52

allotropic forms of carbon 4, 37–8, 85, 114, 224

Amazon Rainforest 120

amides 54

amino acids 114

AMU (unified atomic mass unit) 58

aniline 72

animals
animal chemistry 49–50, 51, 52
and carbon in the biosphere 196–7

anisotropy 90

Ansanto valley, Italy, Mefite lake 11–12, 14–15, 23

anthraxolite 36

the anthropic principle 186–9

the Anthropocene 119, 168, 179, 206–21, 232
arrow of time model 210–13, 215–16
and catastrophism 210, 219
as a grand narrative 208–10, 214
and multiple temporalities 216–21
scalability of 213–16

antiseptic carbons 1

Apollo 17, 20

aquatic environments 204

Aramco 131

archaeological dating 2, 112, 216

Archer, David 217–18

Archimedes 82

Aristotelianism 3, 46–7

Arnold, James 111

aromatic compounds 72

asymmetry in carbon chemistry 63–9, 72
as a disposition 69–70

atomic bombs 111

atomic masses, quantifying 57–9

atomic number 47

atomic weight 43, 44, 47, 58, 59

atomicity 60–1

atoms 6, 36, 59
asymmetric 63–9
dispositions 69–70
and elements 42–3, 44

Augustine 19–20

Auzanneau, Matthieu 131

Avogadro, Amedeo 36, 42, 58, 59

Bachelard, Gaston 48

Bacon, Roger 81–2